Compact Textbooks in Mathematics

This textbook series presents concise introductions to current topics in mathematics and mainly addresses advanced undergraduates and master students. The concept is to offer small books covering subject matter equivalent to 2- or 3-hour lectures or seminars which are also suitable for self-study. The books provide students and teachers with new perspectives and novel approaches. They may feature examples and exercises to illustrate key concepts and applications of the theoretical contents. The series also includes textbooks specifically speaking to the needs of students from other disciplines such as physics, computer science, engineering, life sciences, finance.

- **compact:** small books presenting the relevant knowledge
- **learning made easy:** examples and exercises illustrate the application of the contents
- **useful for lecturers:** each title can serve as basis and guideline for a semester course/lecture/seminar of 2-3 hours per week.

Marek Galewski

Basics of Nonlinear Optimization

Around the Weierstrass Theorem

Marek Galewski
Institute of Mathematics
Łódź University of Technology
Łódź, Poland

ISSN 2296-4568   ISSN 2296-455X   (electronic)
Compact Textbooks in Mathematics
ISBN 978-3-031-77159-0   ISBN 978-3-031-77160-6   (eBook)
https://doi.org/10.1007/978-3-031-77160-6

This book is published under the imprint Birkhäuser, www.birkhauser-science.com by the registered
company Springer Nature Switzerland AG
The registered company address is: Gewerbestrasse 11, 6330 Cham, Switzerland

If disposing of this product, please recycle the paper.

*To my children, Katarzyna and Krzysztof,
who, despite studying textbooks, will likely
not read this one*

Preface

The aim of this textbook is to introduce the reader to various optimization techniques related to the Weierstrass Theorem, which concerns the extrema of continuous functions over compact sets in $\mathbb{R}^N$. We explore several variants of this remarkable result in both finite- and infinite-dimensional contexts, with supporting examples and discussion. Exercises are integrated within the text, rather than placed at the end of each chapter, as they include observations, remarks, and examples that enhance comprehension of the subsequent material. The reader is advised to at least go through the exercises; however, we strongly recommend solving them as they arise. These exercises aim to develop essential skills for future tasks or provide supplementary results. The primary functional spaces studied in this textbook are Sobolev spaces and spaces of continuous functions. Additionally, we include certain background material as it becomes relevant to the main narrative, rather than presenting it in an introductory chapter.

For completeness, we review optimization concepts typically introduced in calculus courses. Our discussion begins with Euclidean space and extends into the infinite-dimensional framework, culminating in the direct variational method. We examine differentiation in infinite-dimensional spaces, introducing relevant frameworks and methods from functional analysis. We assume the reader possesses a basic knowledge of functional analysis, including Hilbert and Banach spaces, Lebesgue integration, and differential equations, for which we provide literature recommendations in the bibliography. When introducing concepts such as weak convergence, we limit the level of generality to what is necessary. Minimizers are primarily studied in reflexive spaces; however, several results regarding minimization in non-reflexive Banach spaces are discussed, particularly for continuous and continuously differentiable functions on the unit interval. We illustrate the technique for finding minimizers in these spaces using the concept of convexity. However, our analysis of the classical calculus of variations is somewhat superficial, as this area currently attracts limited attention. Nevertheless, we believe that understanding optimization within the context of continuous and continuously differentiable functions remains valuable. As a result, when examining the differentiability and continuity of Euler action functions, we compare different spaces in which a functional is evaluated. A similar analysis is conducted when discussing the regularity results related to the du Bois-Reymond Lemma. Several detailed examples are provided to equip the reader with tools for exploring more advanced literature and topics. In our search for minimizers of action functionals and our study of the solvability of associated Dirichlet problems, we mainly focus on ordinary differential equations (ODEs). The justification for our examples and exercises lies primarily in the single-variable context. In the concluding chapter however, we apply the previously discussed approaches to the context of partial differential equations (PDEs). The section on PDEs is rather concise. To simplify matters for the reader, we introduce a basic case involving the open unit ball. Consequently, we refrain from discussing numerous nuanced issues related to boundary conditions and various domain types. On the other hand, we highlight the differences between single-variable and multi-variable scenarios.

These notes were partially used in elective courses on optimization for mathematics students at the Technical University of Lodz between 2011 and 2018. I decided to complete them in the form of a book, which should be regarded as a brief introduction to the topic of nonlinear optimization and variational methods, enabling the reader to study more advanced and comprehensive texts, to which we provide direction at the end. I would like to thank Dr. Michał Bełdziński, my former Ph.D. student, for carefully reviewing this text and pointing out several corrections. I am also grateful for the encouragement and support I received from Dr. Dorothy Mazlum, the Birkhäuser Compact Textbooks Series Editor.

Łódź, Poland   Marek Galewski
September 2024

Contents

The Weierstrass Theorem the Origin of Optimization

1.1   Introductory Remarks

In this chapter, we follow several sources retaining some structure given in our lecture notes [26]. For background in optimization, excellent sources include [9], which provides a thorough introduction to classical optimization techniques, [23], which offers numerous examples and exercises, and [32], which develops optimization techniques in Banach spaces. Moreover, [41] addresses optimization in $\mathbb{R}^N$. For the background in functional analysis, we refer to [8] and also [46].

Let us fix some notation, which will be used later on. Let E be a real Banach space. The space is separable if it contains a dense and countable subset. Let E^* denote the adjoint space, i.e., the space of all linear and continuous mappings defined on E with values in $\mathbb{R}$ which will be called functionals. The norm in E will be denoted by $\|\cdot\|$ and the norm in E^* by $\|\cdot\|_*$ so that to avoid any confusion. We would underline which norm we mean by writing a subscript when necessary. We will use notation $(\cdot, \cdot)_E$ for the scalar product in E in case it is a Hilbert space.

If $f \in E^*$, $x \in E$, then by $\langle f, x \rangle$, we mean a **duality pairing** between E^* and E, i.e. the action of the functional f on the element x. Each element $x \in E$ defines also a linear and continuous functional on E^* via a duality pairing. If $E^{**} = (E^*)^*$ denotes the second dual, we can define a canonical embedding $\chi : E \to E^{**}$. In case such a map is surjective, we say that space E **is reflexive**. In several cases, we shall consider Hilbert spaces, which are necessarily reflexive, but we will also work in Banach spaces that are not reflexive.

The norm in $\mathbb{R}^N$ we will denote as the absolute value, while the scalar product as regular multiplication. As far as differentiation is concerned for the classical derivative we keep notation u', where u is some function, while for an a.e. derivative we will use symbol $\dot{u}$. Sometimes we will use symbol $\frac{d}{dx}$. We tend to denote points by x, x_0, while symbols u, u_0 are reserved for functions.

We are concerned with finding **critical points** (i.e., such points at which a derivative becomes zero) to "energy" functionals $F : E \to \mathbb{R}$ under some differentiability

M. Galewski, *Basics of Nonlinear Optimization*, Compact Textbooks in Mathematics, https://doi.org/10.1007/978-3-031-77160-6_1

assumption, using minimization techniques. Put $\frac{d}{dx}F = f$. Knowing that the critical point exists we immediately see that the following equation

$$f(x) = 0, \tag{1.1}$$

understood in a sense of the space E^*, is solvable. From a classical calculus course, we recall that to find a (local) minimizer of a function, we set the derivative to zero and next apply a second-order test (in case of several variables) or else a first-order test if applicable. In this chapter, we reverse this approach: to solve Eq. (1.1) we will look for minimizers of F. Knowing that a minimizer exists and applying *the Fermat Rule*, we see that solution to (1.1) exists.

By a **minimizer** of a functional $F : E \to \mathbb{R}$ we mean such a point $x_0 \in E$ that

$$F(x) \geq F(x_0) \text{ for all } x \in E.$$

There may exist a limited number of minimizers like for $x \longmapsto x^2$ (unique) and $x \longmapsto (x^2 - 1)^2$ (two), or infinitely many like for $x \longmapsto \sin x$ or no minimizers for $x \longmapsto \exp(x)$. Apart from the global minima, i.e., those that we look for over the whole space, there are local and constrained ones. We say that $x_0 \in E$ is a local minimizer if there is a neighborhood $U(x_0)$ of x_0 such that $F(x) \geq F(x_0)$ for all $x \in U(x_0)$. When the domain of F is some subset D of E, we redefine the aforementioned notions accordingly by taking $U(x_0) \cap D$.

The primary tool used in this text is the well known *Weierstrass Theorem* and its numerous generalizations and corollaries. We begin with considering functionals defined on $\mathbb{R}^N$ which allows us to proceed to the infinite-dimensional case.

The following simple exercises present some interesting observations about finding minima (and maxima) of functions, showing the need for caution. Unless otherwise stated, the functions are assumed to be defined in their natural domains.

Exercise 1.1

Show that a function

$$F(x) = \sin x$$

has infinitely many maxima and minima.

Exercise 1.2

Show that a function

$$F(x) = \frac{1}{1 + x^2}$$

is bounded, has no global minimum, but has a global maximum.

Exercise 1.3

Show that a function

$$F(x) = \arctan x$$

is bounded without extrema.

Exercise 1.4

Show that a function

$$F(x) = (\arctan x)^3$$

is bounded, has no extrema, but has critical points.

Exercise 1.5

Show that a function

$$F(x) = \arctan x \cdot \sin x$$

is bounded and has local minima and maxima, which, however, are not global.

Exercise 1.6

Show that a function

$$F(x) = \begin{cases} x^2 \left(2 + \sin \frac{1}{x}\right), & x \neq 0, \\ 0, & x = 0 \end{cases}$$

has the following properties: it has a local minimizer at $x_0 = 0$, but in any neighborhood around it, the function changes from decreasing to increasing infinitely many times.

Exercise 1.7

Show that a function

$$F(x, y) = x^2 - y^2 + 2 \exp\left(-x^2\right)$$

has a strict local minimum that is not global.

Exercise 1.8

Show that a function

$$F(x, y) = \sin y - x^2$$

has infinitely many local maxima and no local minima.

Exercise 1.9

Show that a function

$$F(x, y) = \left(x - x^2\right)\left(x - 3y^2\right)$$

has a local minimum on every line passing through the point $(0, 0)$, but that point $(0, 0)$ itself is not a local minimizer of F.

1.2   Lower Semicontinuity and the Weierstrass Theorem

In this chapter, we consider functions over $\mathbb{R}^N$, but we will provide relevant notions directly in a Banach space, as they are similarly defined when norm topologies are involved. Analyzing the proof of the Weierstrass Theorem on the existence of extrema for a continuous function over a closed and bounded set (see Theorem 1.1 below) we observe that the continuity assumption may be too strong when looking for a minimum. Consider the example of a discontinuous function

$$F(x) = \begin{cases} x^2, & x \in [-1, 0) \cup (0, 1], \\ -1, & x = 0, \end{cases} \tag{1.2}$$

which has a minimizer $x_0 = 0$ over $[-1, 1]$. While not continuous, the function F is lower semicontinuous, which is sufficient as far as minimization is concerned.

Definition 1.1 (Lower Semicontinuity)

Let E be a real Banach space. A functional $F : E \to \mathbb{R}$ is called lower semicontinuous at $x_0 \in E$ if

$$\liminf_{x \to x_0} F(x) \geq F(x_0) \tag{1.3}$$

which is understood as follows:

$$\forall_{\varepsilon>0}\exists_{\delta>0}\forall_{x\in E} 0 < \|x - x_0\| < \delta \implies F(x) + \varepsilon \geq F(x_0).$$

If condition (1.3) holds for every $x_0 \in E$, then we say that F is *lower semicontinuous on E*.

Continuity in a real Banach space E we understand in a usual manner, i.e., a functional $F : E \to \mathbb{R}$ is continuous at $x_0 \in E$ provided that for every sequence (x_n) converging strongly in E to some x_0 we have

$$\lim_{n \to +\infty} F(x_n) = F(x_0).$$

We have also the standard $\varepsilon - \delta$ definition:

$$\forall_{\varepsilon > 0} \exists_{\delta > 0} \forall_{x \in E} \ \|x - x_0\| < \delta \implies |F(x) - F(x_0)| < \varepsilon.$$

Any continuous functional is obviously lower semicontinuous, but the reverse is not true, as seen in the example of the function given by 1.2).

If we sum a continuous functional $G : E \to \mathbb{R}$ and a lower semicontinuous functional $F : E \to \mathbb{R}$, the result is a lower semicontinuous functional because

$$\liminf_{x \to x_0} (G(x) + F(x)) = \lim_{x \to x_0} G(x) + \liminf_{x \to x_0} F(x).$$

Even if G is only lower semicontinuous, the assertion still holds:

$$\liminf_{x \to x_0} (G(x) + F(x)) \geq \liminf_{x \to x_0} G(x) + \liminf_{x \to x_0} F(x).$$

While checking lower semicontinuity is harder than checking continuity, we have the following useful criterion, for which we introduce some additional notions. Let $F : E \to \mathbb{R} \cup \{-\infty, +\infty\}$. We define the **effective domain** of F as:

$$\mathrm{dom}\, F = \{x \in E : F(x) < +\infty\}.$$

In case $\mathrm{dom}\, F \neq \emptyset$ and $F(x) > -\infty$ for all $x \in E$ we call the function **proper**. Further on, we will mostly consider such functions that are either proper or everywhere finite. **The (Lebesgue) level set** F^α is defined for each $\alpha \in \mathbb{R}$ as follows:

$$F^\alpha = \{x \in E : F(x) \leq \alpha\}$$

We also introduce the notion of **the epigraph** of $F : E \to \mathbb{R}$, defined as the following set:

$$\mathrm{Epi}(F) = \{(x, \alpha) \in E \times \mathbb{R} : F(x) \leq \alpha\}.$$

Lemma 1.1 (Level Sets, Epigraph, and the Lower Semicontinuity)
Let E be a real Banach space. Let $D \subset E$ be a closed set. The following conditions are equivalent:

(a) The functional $F : D \to \mathbb{R}$ is lower semicontinuous.
(b) The set F^α for each $\alpha \in \mathbb{R}$ is closed.
(c) The set $\mathrm{Epi}\,(F)$ jest closed.

Proof Assume that $F : D \to \mathbb{R}$ is lower semicontinuous. Fix $\alpha \in \mathbb{R}$ and take sequence $(x_n) \subset F^\alpha$ that converges in norm in E to some $x_0 \in D$ (that is $\lim_{n \to +\infty} \|x_n - x_0\| = 0$). Then we have:

$$F(x_0) \le \liminf_{n \to +\infty} F(x_n) \le \alpha.$$

This implies that F^α is closed.

Assume now that F^α is closed for each fixed α. Suppose that F is not lower semicontinuous at some $x_0 \in D$. This means that there is a sequence $(x_n) \subset E$ convergent to $x_0 \in D$ and such that

$$\liminf_{n \to +\infty} F(x_n) < F(x_0).$$

Thus there exists some $\alpha_1 \in \mathbb{R}$ for which

$$\liminf_{n \to +\infty} F(x_n) < \alpha_1 < F(x_0)$$

Since (x_n) is convergent, it has a subsequence contained in F^{α_1} that converges to the same x_0. Since F^{α_1} is closed, we have that $F(x_0) \le \alpha_1$, which is impossible. Thus, we have demonstrated that (a) and (b) are equivalent.

Now we show that (a) and (c) are equivalent. Take a sequence $(x_n, \alpha_n) \in \mathrm{Epi}\,(F)$ converging to some (x_0, α_0), which means that $x_n \to x_0$ in E and $\alpha_n \to \alpha_0$ in $\mathbb{R}$. Since $\alpha_n \to \alpha_0$, if we fix $\varepsilon > 0$, then it follows for sufficiently large n that

$$F(x_n) \le \alpha_n \le \alpha_0 + \varepsilon.$$

Since F is lower semicontinuous, we obtain

$$F(x_0) \le \liminf_{n \to +\infty} F(x_n) \le \alpha_0 + \varepsilon.$$

Therefore,

$$F(x_0) \le \alpha_0 + \varepsilon$$

for each $\varepsilon > 0$, which means that $F(x_0) \leq \alpha_0$ and thus $(x_0, \alpha_0) \in$ Epi (F). This now implies that Epi (F) is closed.

Now assume that Epi (F) is closed. Fix an $\alpha \in \mathbb{R}$ such that F^α is nonempty. Then $F^\alpha \times \{\alpha\}$ is closed, which means that F^α is closed. But this implies that F is lower semicontinuous.

▶ **Remark 1.1** It is easy to check that the function $F : \mathbb{R} \to \mathbb{R}$

$$F(x) = \begin{cases} x^2, & x \in [-1, 0) \cup (0, 1], \\ 2, & x = 0 \end{cases}$$

is not lower semicontinuous. It suffice to take $\alpha = 1$ and apply Lemma 1.1. We see that the function F defined as earlier does not have a minimizer. Note that functions that are not lower semicontinuous may still have minimizers, as seen by the example of the following function:

$$F(x) = \begin{cases} x^2, & x \in \left[-1, \frac{1}{2}\right) \cup \left(\frac{1}{2}, 1\right], \\ 2, & x = \frac{1}{2} \end{cases}$$

which is not lower semicontinuous but has a minimizer at $x_0 = 0$. Note, however, that F is lower semicontinuous at the minimizer.

Now we turn the notion of *the minimizing sequence*, which is important to us in the light of the formulation of *the Weierstrass Theorem*. Assume that $D \subseteq E$ and that $F : D \to \mathbb{R}$ is bounded from below (on D). Then F has an infimum over D. Let us recall that for a set $A \subset \mathbb{R}$

$$a := \inf A \Leftrightarrow \begin{cases} \forall_{x \in A} \quad a \leq x, \\ \forall_{\varepsilon > 0} \exists_{x \in A} \quad x \leq a + \varepsilon. \end{cases}$$

Putting

$$a = \inf_{x \in D} F(x) \text{ and } \varepsilon := \frac{1}{n}$$

we see that there exists a sequence $(x_n) \subset D$ such that

$$\liminf_{n \to +\infty} F(x_n) = \inf_{x \in D} F(x). \tag{1.4}$$

Definition 1.2 (Minimizing Sequence)

Let E be a real Banach space. Assume that $D \subseteq E$ and that $F : D \to \mathbb{R}$. Any sequence $(x_n) \subset D$ satisfying (1.4) is called *a minimizing sequence*.

▶ **Remark 1.2** A minimizing sequence (x_n), although not necessarily unique, exists for any functional F, which is bounded from below. It also need not be convergent, even in cases where the limit $\lim_{n \to +\infty} F(x_n)$ exists. For instance, taking

$$F(x) = x^2, \ D = [-3, -1] \cup [1, 3]$$

we see that $x_n = (-1)^n$ forms a divergent minimizing sequence. Note that this sequence has a convergent subsequence. However, if we consider the same function on $D = [-3, 3]$, any minimizing sequence $(x_n) \subset D$ is convergent and consists of almost critical points, i.e., $F'(x_n) \to 0$. Moreover, we can show that for the function

$$F(x) = \begin{cases} x^2 \left(\sin \frac{1}{x} + 1 \right) x \neq 0, \\ 0, \qquad\qquad\qquad x = 0 \end{cases}$$

the point $x_0 = 0$ is a global minimizer, which can be approached by a minimizing sequence consisting of local maxima, as well as by another minimizing sequence consisting of local minima. These examples illustrate that the nature of a minimizing sequence, even for single-variable functions, can be quite complicated. We conclude this remark with an example of a function $F(x) = e^{-x}$ for which $x_n = n$ is a divergent minimizing sequence without any convergent subsequence.

Let us now turn to the finite-dimensional case, i.e., when $E = \mathbb{R}^N$. Then $E = E^*$.

Theorem 1.1 (Weierstrass Theorem for Lower Semicontinuous Function)
Assume that $F : \mathbb{R}^N \to \mathbb{R}$ is lower semicontinuous on D which is a bounded and closed subset of $\mathbb{R}^N$. Then problem

$$\text{find } x_0 \in D \text{ such that } F(x_0) = \inf_{x \in D} F(x) \qquad\qquad (P)$$

has at least one solution.

Proof Observe that F is bounded from below on D. Indeed, otherwise we would find a sequence $(x_n) \subset D$ such that

$$\liminf_{n \to +\infty} F(x_n) = -\infty.$$

Sine D is compact there is a subsequence $\left(x_{k_n}\right) \subset D$ of (x_n) that converges to some x_0. Given that F is lower semicontinuous, we see that

$$-\infty < F(x_0) \leq \liminf_{k_n \to +\infty} F(x_{k_n}) = \liminf_{k_n \to +\infty} F(x_{k_n}) \to -\infty,$$

which implies that $F(x_0) = -\infty$. But this is impossible.

Since F is bounded from below on D, it has its infimum over D. Consequently, there exists a minimizing sequence $(x_n) \subset D$ that, up to a subsequence (x_{k_n}), converges to some $x_0 \in D$. By employing the lower semicontinuity of F and using obvious relations we see that

$$\inf_{x \in D} F(x) \leq F(x_0) \leq \liminf_{k_n \to +\infty} F(x_{n_k}) = \inf_{x \in D} F(x)$$

Therefore $F(x_0) = \inf_{x \in D} F(x)$. This finishes the proof.

The aforementioned theorem can be rephrased as follows: *Let $F : \mathbb{R}^N \to \mathbb{R}$ be a lower semicontinuous functional and let D be a bounded and closed subset of $\mathbb{R}^N$. Then each minimizing sequence (x_n) for functional F over D has a convergent subsequence and there is at least one $x_0 \in D$ such that*

$$\min_{x \in D} F(x) = \inf_{n \in \mathbb{N}} F(x_n) = F(x_0).$$

We will sometimes write $\min_{x \in D}$ instead of $\inf_{x \in D}$ when it is achieved at some point in D. The aforementioned proof works also in infinite dimensional case provided we explicitly assume that D is compact. Under assumptions that it is closed and bounded, we are not able to shift Theorem 1.1, at least with norm convergence, to the infinite-dimensional setting. This is because the closed ball in infinite-dimensional reflexive space is not compact. We recall the following classical example:

Example 1.1

By l^2 we understand a space of sequences $(x_n) \subset \mathbb{R}$ such that

$$\sum_{n=1}^{\infty} x_n^2 < +\infty.$$

We see that l^2 becomes a Hilbert space when endowed with the following scalar product

$$(y, x)_{l^2} = \sum_{i=1}^{\infty} y_i x_i \text{ for } y, x \in l^2.$$

Consider the sequence from the unit sphere of l^2

$$e_1 = (1, 0, 0, \ldots), \; e_2 = (0, 1, 0, 0, ..), \ldots$$

This sequence is bounded, yet it does not contain any strongly convergent subsequence.

From the aforementioned, it follows that the infinite-dimensional case appears much more difficult, and we will tackle it with some additional preparations.

When D is not bounded, for example, when $D = \mathbb{R}^N$, there is also a version concerning the existence of a minimizer for a lower semicontinuous functional.

Definition 1.3 (Coercive Functional)

Let E be a real Banach space. We say that a functional $F : E \to \mathbb{R}$ is *coercive* if

$$\lim_{\|x\| \to +\infty} F(x) = +\infty.$$

Example 1.2

The following coercive function

$$F(x) = \begin{cases} x^2, & x \neq \frac{1}{2}, \\ 4, & x = \frac{1}{2} \end{cases}$$

is not continuous and with a global minimizer.

It is important to note that the notion of coercivity pertains to the behavior of the function at infinity and says nothing about its continuity. We provide a few examples:

Example 1.3

The following functions from $\mathbb{R}^2$ to $\mathbb{R}$ are obviously coercive

$$F(x, y) = \sqrt{x^2 + y^2} = |(x, y)|,$$

$$F(x, y) = x^2 + y^2.$$

On the other hand, linear (affine) functions are never coercive.

Example 1.4

The function $F : \mathbb{R}^2 \to \mathbb{R}$ given by

$$F(x, y) = x^2 + 2xy + y^2$$

is not coercive. Indeed,

$$F(x, y) = (x + y)^2$$

which means that, if $y = -x$ we have $|(x, -x)| \to +\infty$ and $F(x, -x) = 0$.

From the aforementioned example, we note that if F is coercive with respect to each variable separately (with the other held constant), it need not be coercive as a function of two variables.

Example 1.5

Function $F : \mathbb{R}^2 \to \mathbb{R}$ given by

$$F(x, y) = e^{x^2} + x^{100} + e^{y^2} + y^{100}$$

is coercive since (by the repeated use of the de l'Hospital rule)

$$\lim_{t \to +\infty} \frac{e^{t^2}}{t^{100}} = +\infty.$$

Coercivity will aid us in minimizing functionals over unbounded sets. Note that there are functions that are not coercive yet have (unique) minimizers, such as $F(x) = \exp(-x^2)$. Further on, we will examine coercive functions defined on Banach spaces.

Theorem 1.2 (Weierstrass Theorem for Coercive l.s.c. Functional)
Assume that functional $F : \mathbb{R}^N \to \mathbb{R}$ is lower semicontinuous and coercive. Then F has at least one minimizer (in other words, problem (P) has at least one solution with $D = \mathbb{R}^N$).

Proof Let $x_0 \in \mathbb{R}^N$ and put $\alpha_0 = F(x_0)$. Since F is lower semicontinuous, it follows by Lemma 1.1 that the set F^{α_0} is closed. Because the functional F is coercive, there is a number $r > 0$ such that

$$\|x\| > r \Rightarrow F(x) > F(x_0).$$

Therefore F^{α_0} is also bounded. Hence, by Theorem 1.1 there is some $\overline{x} \in F^{\alpha_0}$ such that

$$F(\overline{x}) \leq F(x) \text{ for all } x \in F^{\alpha_0}.$$

Now take any $x \in \mathbb{R}^N$. If $x \in F^{\alpha_0}$, then of course $F(\overline{x}) \leq F(x)$. Otherwise $F(x) > F(x_0) \geq F(\overline{x})$.

We can rephrase the aforementioned theorem in the language of minimizing sequences, noting that for a lower semicontinuous functional defined on a finite-dimensional space, the boundedness of every minimizing sequence implies the existence of a minimizer. Thus, we have obtained a general scheme that will prove useful in what follows.

We conclude our remarks with natural observation concerning minimization of a differentiable functional over $\mathbb{R}^N$ (which is necessarily lower semicontinuous). This result follows from Theorem 1.2 by the application of the Fermat Rule, which renders 0 a derivative at a minimizer or a maximizer.

Theorem 1.3 (Weierstrass Theorem for a Differentiable and Coercive Functional)
Assume that $F : \mathbb{R}^N \to \mathbb{R}$ is differentiable and coercive. Then the functional F has at least one minimizer x_0 such

$$F'(x_0) = 0.$$

The aforementioned theorem directly generalizes to any finite-dimensional space. A standard sufficient condition for differentiability (in a finite dimensional space) is the existence and continuity of all first-order partial derivatives. We note the following version of Theorem 1.3:

Theorem 1.4
Assume that $F : \mathbb{R}^N \to \mathbb{R}$ is lower semicontinuous and coercive. Assume it has first-order partial derivatives with respect to all variables. Then functional F has at least one minimizer x_0 such

$$F'_{x_i}(x_0) = 0 \text{ for } i = 1, 2, \ldots, N.$$

We had to add the assumption about lower semicontinuity for the reasons outlined in the following example:

Example 1.6

The function

$$F(x, y) = \begin{cases} -1, & x = y^2, y > 0, \\ 0, & \text{otherwise} \end{cases}$$

is not continuos at $(0, 0)$, while it has both partial derivatives equal to 0.

We mention also some special case of Theorem 1.2 aimed at minimization of a function over closed and unbounded sets, leaving the immediate proof as an exercise:

Theorem 1.5
Let $D \subset \mathbb{R}^N$ be a closed and unbounded set. Assume that functional $F : D \to \mathbb{R}$ is lower semicontinuous and such that $F(x_n) \to +\infty$ for any sequence $(x_n) \subset D$ with $|x_n| \to +\infty$. Then F has at least one minimizer over D.

It should be noted that with the help of Theorem 1.5 we obtain a minimizer, which need not be a critical point.

Example 1.7

Let $D = \{x \in \mathbb{R}^N : |x| \geq 1\}$ and put $F(x) = |x|^2$. Then F satisfies the coercivity condition given in Theorem 1.5. The arguments of a minimum cover the unit sphere, but neither is the critical point.

1.3 Applications to Minimization Problems

We proceed with some simple calculus applications of Theorem 1.3. These can be skipped by more advanced readers. We recall that, concerning the global (or local) minimum, it is important to distinguish between the minimal value and the infimal value. More precisely, let $F : \mathbb{R}^2 \to \mathbb{R}$ be $F(x, y) = x^2 + y^2$. Then

$$\min_{(x,y)\in\mathbb{R}^2} F(x, y) = \inf_{(x,y)\in\mathbb{R}^2} F(x, y) = F(0, 0).$$

When $F : \mathbb{R}^2 \to \mathbb{R}$ is defined by $F(x) = e^{x+y}$, then

$$\inf_{(x,y)\in\mathbb{R}^2} F(x, y) = 0$$

and there is no minimizer.

Example 1.8

Let us consider the following C^2 function $F : \mathbb{R}^2 \to \mathbb{R}$

$$F(x, y) = x^4 - 4xy + y^4$$

which we aim to minimize over $\mathbb{R}^2$. We find that

$$\nabla F\left(x, y\right) = \left(4x^3 - 4y, 4y^3 - 4x\right) \text{ and } Hf\left(x, y\right) = \begin{bmatrix} 12x^2 & -4 \\ -4 & 12y^2 \end{bmatrix}.$$

We see that $\det Hf\left(\frac{1}{2}, \frac{1}{2}\right) < 0$, so the second-order test fails here. Nevertheless, the function F satisfies:

$$F\left(x, y\right) = x^4 - 4xy + y^4 = x^4 + y^4\left(1 - \frac{4xy}{y^4}\right) \text{ or } F\left(x, y\right) = x^4 - 4xy + y^4$$

$$= y^4 + x^4\left(1 - \frac{4xy}{x^4}\right)$$

for any (x, y) away from $(0, 0)$, which implies that it is coercive. The critical points are:

$$(0, 0), \ (1, 1), \ (-1, -1)$$

and we see that:

$$F\left(0, 0\right) = 0, \ F\left(1, 1\right) = F\left(-1, -1\right) = -2$$

which means that there are two (distinct) global minimizers.

Example 1.9

Let us consider the following C^2 function $F : \mathbb{R}^2 \to \mathbb{R}$

$$F\left(x, y\right) = x^3 - 12xy + 8y^3.$$

By a direct computation we have for all $(x, y) \in \mathbb{R}^2$

$$\nabla F\left(x, y\right) = \left(3x^2 - 12y, 24y^2 - 12x\right) \text{ and } Hf\left(x, y\right) = \begin{bmatrix} 6x & -12 \\ -12 & 48y \end{bmatrix}.$$

In order to find critical points, we solve the system

$$\begin{cases} 3x^2 - 12y = 0, \\ -12x + 24y^2 = 0. \end{cases}$$

Multiplying the first equation by x and the second by y, which provide $x = 2y$ and we find the critical points: $(0, 0)$, $(1, 2)$. The Hesse matrix is positive definite at $(1, 2)$, which implies that this is a strict local minimizer. We see that $F\left(0, 0\right) = 0$ and that at $(0, 0)$ the Hesse matrix is neither positive nor negative definite. Moreover, the function

$$g(x) = F(x, 0) = x^3$$

has both positive and negative values in any neighborhood of $(0, 0)$. Thus there is no extremum there. The analysis of the function g defined earlier allows us to conclude that point $(1, 2)$ is not a global minimizer since

$$\lim_{x \to -\infty} g(x) = -\infty \text{ and } \lim_{x \to +\infty} g(x) = +\infty.$$

Exercise 1.10

Check that both functions

(a) $F_1(x, y) = x^4 + y^3$,
(b) $F_2(x, y) = x^2 + y^4$

have critical point at $(0, 0)$, which is a global minimizer for F_2 only. Note that the Hesse matrices are positive semidefinite there.

Exercise 1.11

Check that function

$$F(x, y) = e^x + e^y + 2e^{-x-y}$$

has positive definite Hesse matrix at every point. Find the global minimizer.

Exercise 1.12

Check which of the following functions are coercive:

(a) $F_1(x, y) = x^4 + y^4 - 3xy - x^2 - y^2$,
(b) $F_2(x, y) = x^3 + y^3 + xy$,
(c) $F_3(x, y) = x^4 + y^4 - 3xy^2$,
(d) $F_4(x, y) = x^4 + y^4 - 3xy^3$.

Exercise 1.13

Find global extrema of the following functions:

(a) $F_1(x, y) = x^2 - 4x + 2y^2 + 7$,
(b) $F_2(x, y) = e^{-x^2 - y^2}$,
(c) $F_3(x, y) = (x - y)^2 + (y - 1)^2$,
(d) $F_4(x, y) = x^4 + 16xy + y^8$.

Exercise 1.14

Consider function

$$F(x, y) = x^4 + y^4 - 32y^2.$$

(a) Find point at which the Hesse matrix is indefinite.
(b) Show that F is coercive.
(c) Find all global minimizers.

We conclude this section with a direct application of minimization techniques to the least square optimization allowing for fitting the data into the straight line (consult [41] for a more abstract treatment):

Exercise 1.15 (Least Square Optimization)

Given a collection of points $(t_1, s_1), (t_2, s_2), \ldots, (t_n, s_n)$ find a straight line, i.e., a function

$$r(t) = b + at$$

whose graph is lying as close as possible to all of these points. This means that we need to minimize over $\mathbb{R}^2$ the function

$$(a, b) \longmapsto (r(t_1) - s_1)^2 + (r(t_2) - s_2)^2 + \ldots + (r(t_n) - s_n)^2.$$

Prove that given

$$\bar{t} = \frac{1}{n} \sum_{i=1}^{n} t_i, \quad \bar{s} = \frac{1}{n} \sum_{i=1}^{n} s_i$$

the values for a and b are as follows:

$$a = \frac{\frac{1}{n} \sum_{i=1}^{n} t_i s_i - \bar{t}\bar{s}}{\frac{1}{n} \sum_{i=1}^{n} t_i^2 - (\bar{t})^2} \text{ and } b = \bar{s} - a\bar{t}.$$

1.4   Applications to Global Invertibility

We start this section by recalling that a C^1-mapping $f : \mathbb{R}^N \to \mathbb{R}^N$ is locally invertible and its inverse function is also C^1 provided that $\det f'(x) \neq 0$ for any $x \in \mathbb{R}^N$. The last assumption means that $f'(x)$ is invertible for any $x \in \mathbb{R}^N$, i.e. $f'(x)$ is a linear mapping, which is a bijection between $\mathbb{R}^N$ and $\mathbb{R}^N$. Let us first consider C^1-functions $f : \mathbb{R} \to \mathbb{R}$ such that $f'(x) \neq 0$ for any $x \in \mathbb{R}$. Of course, such a function need not be globally invertible as a simple example of arctan shows. On the other hand, a C^1-function $f_1(x) = x^3$ is globally invertible while its inverse is not C^1. This function is not locally C^1-invertible around 0. At the same time, function $f_2(x) = x^3 + x$ is locally (and globally) invertible. We see that $|f_i(x)| \to +\infty$ as $|x| \to +\infty$ for $i = 1, 2$. In fact, it is easy to solve the following exercise:

Exercise 1.16

Prove that a locally invertible coercive C^1-function $f : \mathbb{R} \to \mathbb{R}$ is globally invertible, that is $f^{-1} : \mathbb{R} \to \mathbb{R}$ is defined and C^1. Hint: Consider an auxiliary coercive C^1 functional $g : \mathbb{R} \to \mathbb{R}$ defined for a fixed $y \in \mathbb{R}$ by

$$g(x) = |f(x) - y|^2.$$

Show that functional g has a minimizer $\overline{x}$,, which by Fermat's rule, satisfies equation $(f(\overline{x}) - y)f'(\overline{x}) = 0$. Now draw conclusion about surjectivity. To show that f is injective assume that for some y there exist distinct x_1, x_2 such that $f(x_1) = f(x_2) = y$ and apply the Rolle Theorem.

We acknowledge that the argument given in the hint is overly sophisticated for a single-variable setting since, under the given assumptions, f is strictly monotone. However, this proof has the advantage of a generalization to maps between Euclidean spaces known as the Hadamard Theorem, see [33, Theorem 5.4]. Of course, Rolle's Theorem cannot be used in the multidimensional setting and instead the following three critical points theorem is applied:

Theorem 1.6 (Courant Theorem)
Suppose that a C^1-functional $f : \mathbb{R}^N \to \mathbb{R}$ is coercive and possesses two distinct strict relative minima x_1 and x_2. Then f possesses a third critical point x_3, distinct from x_1 and x_2, which is not a relative minimizer, that is, in every neighborhood of x_3, there exists a point x such that $f(x) < f(x_3)$.

As a direct application of the Courant Theorem, we recall Example 1.8. A continuously differentiable map $f : \mathbb{R}^N \to \mathbb{R}^N$ is called a **diffeomorphism** if it is a bijection and its inverse $f^{-1} : \mathbb{R}^N \to \mathbb{R}^N$ is continuously differentiable as well. Recalling *the Inverse Function Theorem*, a continuously differentiable mapping $f : \mathbb{R}^N \to \mathbb{R}^N$ such that $f'(x) \in \text{Isom}(\mathbb{R}^N, \mathbb{R}^N)$ for any $x \in \mathbb{R}^N$ (equivalently $\det f'(x) \neq 0$) defines a **local diffeomorphism**, i.e., for each $x \in \mathbb{R}^N$ there exists an open set U containing x, such that $f(U)$ is open in Y and $f|_U : U \to f(U)$ is a diffeomorphism. If f is a diffeomorphism, it obviously defines a local diffeomorphism. Thus, the main problem to be overcome is to make a local diffeomorphism into a global one. Here we present the approach applicable in a finite-dimensional setting and which is provided by the Hadamard Theorem. In the infinite-dimensional space, one may wish to consult [2] for global invertibility methods.

Theorem 1.7 (Hadamard Theorem)
Assume that $f : \mathbb{R}^N \to \mathbb{R}^N$ is a C^1-mapping such that

(a) $f'(x) \in \mathrm{Isom}(\mathbb{R}^N, \mathbb{R}^N)$ for any $x \in \mathbb{R}^N$,
(b) $|f(x)| \to +\infty$ as $|x| \to +\infty$.

Then f is a diffeomorphism.

Proof By (a) we see that f defines a local diffeomorphism. Thus, it is sufficient to show that f is onto and one-to-one.

First, we show that f is onto. Let us fix any point $y \in \mathbb{R}^N$ and define functional $\varphi : \mathbb{R}^N \to \mathbb{R}$ as follows

$$\varphi(x) = \frac{1}{2}|f(x) - y|^2. \tag{1.5}$$

Observe that φ is a composition of two C^1–mappings, thus $\varphi \in C^1(\mathbb{R}^N, \mathbb{R})$. Moreover, φ is coercive due to its definition and condition (b). It now follows from Theorem 1.3 that there exists at least one minimizer of φ which we denote by $\overline{x}$. We see by *the Chain Rule* and by *the Fermat Rule* that

$$\varphi'(\overline{x}) = f'(\overline{x})(f(\overline{x}) - y) = 0. \tag{1.6}$$

Since by (a) mapping $f'(\overline{x})$ is invertible, we get $f(\overline{x}) - y = 0$. Thus, f is surjective.

Now we argue by contradiction that f is one-to-one. Suppose there are $x_1, x_2 \in \mathbb{R}^N$, $x_1 \neq x_2$, such that $f(x_1) = f(x_2)$. We will apply Theorem 1.6. Thus we put $\zeta = x_1 - x_2$ and define functional $\psi : \mathbb{R}^N \to \mathbb{R}$ by the formula

$$\psi(x) = \frac{1}{2}|f(x + x_2) - f(x_1)|^2. \tag{1.7}$$

Functional ψ is again C^1 and coercive. Next, we see by a direct calculation that $\psi(\zeta) = \psi(0) = 0$. Moreover, 0 is a strict local minimum of ψ, since otherwise, in any neighborhood of 0 we would have a nonzero x with $f(x + x_2) = f(x_1)$ and this would contradict the fact that f defines a local diffeomorphism. Same holds true for element ζ. Thus, by Theorem 1.6 we note that ψ has a critical point v such that $\psi(v) > 0$. Since v is a critical point we have

$$\psi'(v) = f'(v + x_2)(f(v + x_2) - f(x_1)) = 0.$$

Since $f'(v + x_2)$ is invertible, we see that $f(v + x_2) = f(x_1)$. But this means that $\psi(v) = 0$, which is impossible.

Exercise 1.17

Prove that functionals φ and ψ defined by formulas (1.5) and (1.7) are C^1 and coercive. Verify that formula (1.6) holds.

Exercise 1.18

Let E be finite-dimensional space and assume that $F : E \to \mathbb{R}$ is coercive. Prove that F is coercive with respect to any other norm employed on E.

We extend an example from [25] in order to demonstrate the applicability of the Hadamard Theorem:

Example 1.10

We consider the following mapping $f_a : \mathbb{R}^2 \to \mathbb{R}^2$

$$f_a (x, y) = (x + ax, x^3 + y)$$

for any fixed parameter $a \in (-1, 1)$. We apply *the Hadamard Theorem* in order to show that it defines the global diffeomorphism. For this purpose, we shall show that:

(a) *for each $a \in (-1, 1)$ and any $(x, y) \in \mathbb{R}^2$ matrix $f_a'(x, y)$ is nonsingular;*
(b) $|f_a (x, y)| \to +\infty$ *as* $|(x, y)| \to +\infty$.

Let us observe that

$$f_a'(x, y) = \begin{bmatrix} 1 + a & 0 \\ 3x^2 & 1 \end{bmatrix},$$

which implies that $f_a'(x, y)$ is nonsingular for any $(x, y) \in \mathbb{R}^2$. It remains to show that for some suitably selected norm $\|\cdot\|$ functional $(x, y) \to \|f_a(x, y)\|$ is coercive. Thus, we will apply the l^1 norm on $\mathbb{R}^2$. For any $a \in (-1, 1)$ we denote

$$u_a (x, y) = \| f_a (x, y)\|_1 = |x + ax| + \left|x^3 + y\right|, \quad (x, y) \in \mathbb{R}^2.$$

We observe that

$$u_a (x, y) \geq |x| - |a| \cdot |x| + \left|x^3 + y\right| \geq$$
$$(1 - |a|) \left(|x| + \left|x^3 + y\right|\right) = (1 - |a|)\, u_0 (x, y)$$

Denote by B_k the $k-$th quadrant of $\mathbb{R}^2$, that is:

$$B_1 = \left\{ (x, y) \in \mathbb{R}^2 : x \geq 0, y \geq 0 \right\},$$

$$B_2 = \left\{ (x, y) \in \mathbb{R}^2 : x \leq 0, y \leq 0 \right\},$$

$$B_3 = \left\{ (x, y) \in \mathbb{R}^2 : x \geq 0, y \leq 0 \right\},$$

$$B_4 = \left\{ (x, y) \in \mathbb{R}^2 : x \leq 0, y \geq 0 \right\},$$

$$B_{31} = \left\{ (x, y) \in \mathbb{R}^2 : x \geq -\sqrt[3]{y} \geq 0, x \geq 1 \right\},$$

$$B_{32} = \left\{ (x, y) \in \mathbb{R}^2 : 1 \leq x \leq -\sqrt[3]{y} \right\},$$

$$B_{33} = \left\{ (x, y) \in \mathbb{R}^2 : 0 \leq x \leq 1 \right\}.$$

Note that

$$B_3 \subset B_{31} \cup B_{32} \cup B_{33}$$

and also that

$$\bigcup_{i=1}^{4} B_i = \mathbb{R}^2.$$

If $(x, y) \in B_1$ then

$$u_0 (x, y) = x + x^3 + y \geq x + y = |x| + |y|.$$

Moreover, observe that

$$u_a (-x, -y) = u_{-a} (x, y)$$

which implies

$$u_a (x, y) \geq (1 - |a|) u_0 (x, y) \geq (1 - |a|) (|x| + |y|)$$

for each $(x, y) \in B_1 \cup B_2$.

If $(x, y) \in B_{31}$ then

$$u_0(x, y) = x + x^3 + y \geq x = \tfrac{1}{2}|x| + \tfrac{1}{2}|x| \geq$$

$$\tfrac{1}{2}|x| + \tfrac{1}{2}\sqrt[3]{|y|} \geq \tfrac{1}{2}\sqrt[3]{|x| + |y|}$$

For $(x, y) \in B_{32}$ we consider function

$$\varphi(t) = t - t^3, t \in [1, +\infty).$$

We see that φ is nondecreasing and

$$\varphi(x) \geq \varphi\left(-\sqrt[3]{y}\right) = -\sqrt[3]{y} + y.$$

Then

$$u_0(x, y) = x - x^3 - y = \varphi(x) - y \geq -\sqrt[3]{y} + y - y =$$

$$\tfrac{1}{2}\sqrt[3]{|y|} + \tfrac{1}{2}\sqrt[3]{|y|} \geq \tfrac{1}{2}|x| + \tfrac{1}{2}\sqrt[3]{|y|} \geq \tfrac{1}{2}\sqrt[3]{|x| + |y|}.$$

If $(x, y) \in B_{33}$ we see that

$$u_0(x, y) \geq |x| + |y| - |x|^3 \geq |x| + |y| - 1.$$

The aforementioned calculations have been performed for $\alpha = 0$. For any $\alpha \neq 0$ we proceed as above with some technical changes. Summing up the above obtained estimations, we reach the required coercivity, which implies that f_a defines a global diffeomorphism for each $a \in (-1, 1)$.

1.5   Applications to Algebraic Equations

The theory of difference equations, or in other words, discrete equations, is thoroughly described in many excellent sources, among which we suggest [34]. We will focus on the so-called algebraic equation, which allows us to bypass certain details from the theory of difference equations that are not directly relevant to our illustration. For more information on such equations, an interested reader may consult [3] and also [38].

We apply *the Hadamard Theorem* to investigate the unique solvability of the following algebraic equation

$$Ax = f(x), \tag{1.8}$$

where A is some nonsingular $N \times N$−matrix with real entries and $f : \mathbb{R}^N \to \mathbb{R}^N$ is a C^1 mapping. Let A^T be the transpose of A. Then $A^T A$ is positively definite, i.e., it has N distinct eigenvalues ordered as follows

$$0 < \lambda_1 \leq \cdots \leq \lambda_N.$$

Matrix A can be obtained through the discretization of the second-order ordinary differential operator with Dirichlet boundary conditions. See, for example, the mentioned book [34] and also [21]. One can also consult papers [6] and [24] for some results in this direction.

Theorem 1.8

Assume A is nonsingular and that $f : \mathbb{R}^N \to \mathbb{R}^N$ is C^1. Moreover assume that:

(a) there is a constant $0 < a < \sqrt{\lambda_1}$ such that $|f(x)| \leq a|x|$ for all $x \in \mathbb{R}^N$,
(b) $\det(A - f'(x)) \neq 0$ for all $x \in \mathbb{R}^N$.

Then Eq. (1.8) has exactly one solution.

Proof We will apply Theorem 1.7, which means that it suffice to show the coercivity of $x \to |\varphi(x)|$ over $\mathbb{R}^N$, where $\varphi : \mathbb{R}^N \to \mathbb{R}^N$ is given by

$$\varphi(x) = Ax - f(x).$$

We see that for any $x \in \mathbb{R}^N$ we have

$$|\varphi(x)| = |Ax - f(x)| \geq |Ax| - |f(x)|$$
$$\geq \sqrt{(A^T Ax, x)} - a|x| \geq \left(\sqrt{\lambda_1} - a\right)|x|.$$

This leads to the conclusion.

Exercise 1.19

Show that the assertion of Theorem 1.8 holds when we replace assumption (a) with the following: there are constants $0 < a < \sqrt{\lambda_1}$ and $b \geq 0$, $R > 0$ such that $|f(x)| \leq a|x| + b$ for all $x \in \mathbb{R}^N$ with $|x| \geq R$.

Exercise 1.20

Show that the assertion of Theorem 1.8 holds when we replace assumption (a) with the following: there is a constant $0 < a < \sqrt{\lambda_1}$ such that

$$\limsup_{|x| \to +\infty} \frac{f(x)}{|x|} \leq a.$$

Since we are working in a finite-dimensional space, we can obtain a parallel result:

Theorem 1.9

Assume A is nonsingular and that $f : \mathbb{R}^N \to \mathbb{R}^N$ is C^1. Moreover, we assume that:

(a) there is a constant $b > \sqrt{\lambda_N}$ such that $|f(x)| \geq b|x|$ for all $x \in \mathbb{R}^N$,
(b) $\det(A - f'(x)) \neq 0$ for all $x \in \mathbb{R}^N$.

Then Eq. (1.8) has exactly one solution.

Proof Now we see that for $x \in \mathbb{R}^N$

$$|\varphi(x)| \geq |f(x)| - |Ax|$$
$$\geq b|x| - \sqrt{(A^T Ax, x)} \geq (b - \sqrt{\lambda_N})|x|.$$

Hence, Theorem 1.7 can be applied.

Example 1.11

Consider the following nonsingular matrix

$$A = \begin{bmatrix} -2 & 5 \\ 1 & -3 \end{bmatrix}$$

and a mapping $f : \mathbb{R}^2 \to \mathbb{R}^2$ defined by

$$f(x, y) = \left(x^3 + y, 4x + y^3 + 2y \right).$$

The space $\mathbb{R}^2$ is considered with the Euclidean norm and therefore:

$$|(x, y)| \leq 2^{\frac{1}{3}} \sqrt[6]{x^6 + y^6} \text{ for all } (x, y) \in \mathbb{R}^2.$$

Since

$$f(x, y) = \left(x^3, y^3 \right) + (0, 4x) + (y, 2y),$$

we see that

$$|f(x, y)| \geq |(x^3, y^3)| - |(0, 4x)| - |(y, 2y)| = \sqrt{x^6 + y^6} - 4 \cdot |x| - \sqrt{5}|y|$$
$$\geq \tfrac{1}{2}|(x, y)|^3 - 4(|x| + |y|) \geq \tfrac{1}{2}|(x, y)|^3 - 4\sqrt{2}|(x, y)|.$$

Put

$$\varphi(x, y) = f(x, y) - A(x, y), \quad (x, y) \in \mathbb{R}^2.$$

Let $\|A\|$ be a standard matrix norm. Observe that

$$\begin{aligned} |\varphi(x, y)| &\geq |f(x, y)| - |A(x, y)| \\ &\geq \tfrac{1}{2}|(x, y)|^3 - 4\sqrt{2}|(x, y)| - \|A\| |(x, y)| \\ &= |(x, y)| \left(\tfrac{1}{2}|(x, y)|^2 - \left(4\sqrt{2} + \|A\| \right) \right). \end{aligned}$$

Therefore, $(x, y) \to |\varphi(x, y)|$ is coercive. We compute directly that for $(x, y) \in \mathbb{R}^2$

$$f'(x, y) - A = \begin{bmatrix} 3x^2 + 2 & 0 \\ 0 & 3y^2 + 5 \end{bmatrix} \text{ and } \det\left(f'(x, y) - A \right) > 0.$$

Therefore, Theorem 1.7 leads to the conclusion that system

$$\begin{cases} -x^3 - 2x + 4y = 0, \\ -y^3 - 5y - 3x = 0 \end{cases}$$

is uniquely solvable. By a direct calculation we see that $(0, 0)$ solves the aforementioned system. The information that we get is therefore about the uniqueness in this case.

1.6   The Lagrange Multiplier Rule

The Lagrange Multiplier Theorem is concerned with finding constrained extrema. We will present a version of this celebrated result, accompanied by a recently simplified proof in the finite-dimensional case, as detailed in [29]. The proof uses the Weierstrass Theorem applied to an unconstrained penalized problem. Let N, m be given integers.

Theorem 1.10 (Lagrange Multiplier Rule)
Assume that $f : \mathbb{R}^N \rightarrow \mathbb{R}$ *and* $h_i : \mathbb{R}^N \rightarrow \mathbb{R}$, $i = 1, 2, \ldots, m$ *are continuously differentiable and* $\bar{x}$ *is a local minimizer for* f *on the set*

$$S := \{x \in \mathbb{R}^N : h_i(x) = 0, \ i = 1, 2, \ldots, m\}. \tag{1.9}$$

Assume that the vectors $\nabla h_1(\bar{x}), \ldots, \nabla h_m(\bar{x})$ *are linearly independent. Then there exist Lagrange Multipliers* $\bar{\lambda}_1, \ldots, \bar{\lambda}_m \in \mathbb{R}$ *satisfying*

$$\nabla f(\bar{x}) + \sum_{i=1}^{m} \bar{\lambda}_i \nabla h_i(\bar{x}) = 0.$$

Proof Let us select $\delta > 0$ so that $\bar{x}$ is a global minimizer over $\overline{B}(\bar{x}, \delta) \cap S$, where $\overline{B}(\bar{x}, \delta)$ is the closed ball centred at $\bar{x}$ with radius δ. Let us consider for each $k \in \mathbb{N}$ function

$$\phi_k(x) = f(x) + \frac{1}{2}|x - \bar{x}|^2 + \frac{k}{2}|h(x)|^2$$

and the following problem

$$\underset{x \in B(\bar{x}, \delta)}{\text{Minimize}} \ \phi_k(x).$$

Since function ϕ_k is continuous and the constraint set is compact, we apply the Weierstrass Theorem in order to obtain for each $k \in \mathbb{N}$ at least one solution which we denote by x^k.

We shall show that $x^k \rightarrow \bar{x}$. Since (x^k) is necessarily bounded, we have that (x^k) converges up to a subsequence, which we do not renumber, to some limit point x^*. We will examine the properties of this limit. First, note that for all $k \in \mathbb{N}$,

$$f(x^k) + \frac{1}{2}|x^k - \bar{x}|^2 \leq \phi_k(x^k) \leq \phi_k(\bar{x}) = f(\bar{x}). \tag{1.10}$$

The aforementioned means that the sequence $(\phi_k(x^k))$ is bounded from above. Since by the continuity we have $h(x^k) \rightarrow h(x^*)$ the only possibility is that $h(x^*) = 0$. Taking a limit in (1.10) we arrive at the following

$$f(x^*) + \frac{1}{2}|x^* - \bar{x}|^2 \leq f(\bar{x}) \leq f(x^*)$$

which implies that $x^* = \bar{x}$. This shows that the limit point x^* of (x^k) is unique and hence the whole sequence (x^k) converges to $\bar{x}$.

By the definition of the limit for $k \in \mathbb{N}$ large enough, we must have $|x^k - \bar{x}| < \delta$ and hence, x^k locally minimizes ϕ_k over $B(\bar{x}, \delta)$ as well. Therefore, by *the Fermat Rule* $\nabla \phi_k(x^k) = 0$. Calculating $\nabla \phi_k$ directly we obtain

$$\nabla f\left(x^k\right) + x^k - \bar{x} + \sum_{i=1}^{m} \lambda_i^k \nabla h_i\left(x^k\right) = 0 \tag{1.11}$$

with

$$\lambda_i^k := k h_i\left(x^k\right), \ i = 1, \ldots, m.$$

Let $\lambda^k := (\lambda_1^k, \ldots, \lambda_m^k) \in \mathbb{R}^m$. We shall show that (λ^k) is bounded. If this is not the case, i.e., if $|\lambda^k| \to +\infty$, we see that, up to a subsequence, $\frac{\lambda^k}{|\lambda^k|} \to \alpha = (\alpha_1, \ldots, \alpha_m) \in \mathbb{R}^m$. Clearly, $\alpha \neq 0$, since it is the limit of vectors of length one. Thus, dividing (1.11) by $|\lambda^k|$ and taking the limit, upon taking a subsequence which we again do not renumber, by the continuity of the gradients we arrive at $\sum_{i=1}^{m} \alpha_i \nabla h_i(\bar{x}) = 0$. However this contradicts our assumption about the gradients being independent. This results that (λ^k) is bounded. Thus, considering a subsequence that converges to some $\lambda \in \mathbb{R}^m$ and taking the corresponding limit in (1.11) we arrive at the result.

▶ **Remark 1.3** The assumption that the vectors $\nabla h_1(\bar{x}), \ldots, \nabla h_m(\bar{x})$ are linearly independent is a type of the so-called constraint qualification and is used in order to have a multiplier next to f equal 1 (in case of minimization, in case of maximization we put -1). There are another types of constraint qualifications imposed on the constraints. The interested Reader may consult for example the recent paper [9] for some bibliography suggestions. We also note that the sign of a multiplier is irrelevant in presence of equality constraints.

We will formulate also the following version of Lagrange Multiplier Rule, which is very suitable for practical applications, see [4]:

Theorem 1.11 (Lagrange Multiplier Rule-General Version)
Assume that $f : \mathbb{R}^N \to \mathbb{R}$ and $h_i : \mathbb{R}^N \to \mathbb{R}$, $i = 1, 2, \ldots, m$ are continuously differentiable and $\bar{x}$ is a local minimizer for f on the set S given by (1.9). Then for suitable numbers (Lagrange multipliers) λ_i, $0 \leq i \leq m$, not all equal to zero, the Lagrange vector equation holds

$$\lambda_0 \nabla f(\bar{x}) + \lambda_1 \nabla h_1(\bar{x}) + \cdots + \lambda_m \nabla h_m(\bar{x}) = 0. \tag{1.12}$$

We would like to mention an example concerning the case when $\lambda_0 = 0$, i.e., the situation when no constraint qualification is satisfied.

Example 1.12

Consider the problem of minimizing

$$f(x, y) = x$$

subject to

$$h(x, y) = x^3 - y^2 = 0.$$

We see from the constraint set $S = \{(x, y) \in \mathbb{R}^2 : h(x, y) = 0\}$ that $x \geq 0$. Therefore, point $(x_0, y_0) = (0, 0)$ stands for the minimizer of the problem under consideration. When we try to apply the Lagrange Multiplier Rule, we arrive at the following system

$$\begin{cases} \lambda_0 + 3\lambda x_0^2 = 0 \text{ and } 2\lambda y_0 = 0, \\ x_0^3 - y_0^2 = 0. \end{cases}$$

We immediately see that $\lambda_0 = 0$.

We follow with some examples and exercises mainly taken from [4] thereby showing how to apply the Lagrange Multiplier Rule in practice:

Example 1.13

We will find extrema of a function

$$f(x, y) = x^2 + y^2$$

subject to

$$h(x, y) = x^4 + y^4 - 1 = 0.$$

We see that the constraint set $S = \{x \in \mathbb{R}^2 : h(x, y) = 0\}$ is compact, so by the Weierstrass Theorem, we arrive at the conclusion that f has extrema over S. Hence, there is some point (x_0, y_0) which is a minimizer of f over S and which satisfies (1.12) for some non-zero vector (λ_0, λ). Putting $\lambda_0 = 0$ in (1.12) there must exist $\lambda \neq 0$ such that

$$\begin{cases} \lambda x^3 = 0 \text{ and } \lambda y^3 = 0, \\ x^4 + y^4 = 1. \end{cases}$$

Therefore, $\lambda_0 \neq 0$ and so we put $\lambda_0 = 1$ and we write (1.12) as

$$\begin{cases} x + \lambda x^3 = 0 \text{ and } y + \lambda y^3 = 0 \\ x^4 + y^4 = 1. \end{cases}$$

Then either $x = 0$ in which case $y = \pm 1$, or $y = 0$ in which case $x \pm 1$, or else $x \neq 0$ and $y \neq 0$ which implies that $x = y = -\frac{1}{\lambda}$, i.e. $|x| = |y| = 2^{-1/4}$. Having calculated the values of f at all points obtained we see that at $(\pm 1, 0)$, $(0, \pm 1)$ function f has a minimum over S whose value is 1.

Exercise 1.21

Find the global maxima in Example 1.13.

Exercise 1.22

Find the extrema of

$$f(x, y) = x^2 + y^2$$

subject to condition

$$3x + 4y = 1.$$

Exercise 1.23

Find the extrema of

$$f(x, y) = 3x + 4y$$

subject to condition

$$x^2 + y^2 = 1.$$

Exercise 1.24

Find the extrema of

$$f(x, y, z) = xy^2z^3$$

subject to condition

$$x + y + z = 1.$$

Exercise 1.25

Find the extrema of

$$f(x, y, z) = xyz$$

subject to conditions

$$x + y + z = 0 \text{ and } x^2 + y^2 + z^2 = 1.$$

Exercise 1.26

Find the minimum of a linear functional defined on $\mathbb{R}^N$ over a unit sphere.

Now we turn to examination of extrema of *the Raleigh Quotient* (here we use any norm generated by the associated scalar product and this is why we use some general notation):

Definition 1.4

Let A be a symmetric $N \times N$−matrix. *The Raleigh Quotient* is the function $R : \mathbb{R}^N \setminus \{0\} \to \mathbb{R}$

$$R(x) = \frac{\langle x, Ax \rangle}{\|x\|^2}, \qquad \text{for } x \neq 0.$$

Observe that

$$R(x) = \left\langle \frac{x}{\|x\|}, A\frac{x}{\|x\|} \right\rangle = \langle y, Ay \rangle, \qquad \text{where } y = \frac{x}{\|x\|}.$$

When we minimize or maximize the function R, it suffice to apply *the Lagrange Multiplier Rule* in finding the extrema of $x \longmapsto \langle x, Ax \rangle$ on a unit sphere.

Theorem 1.12

Assume that A is symmetric and positive definite $N \times N$−matrix with eigenvalues ordered as

$$\lambda_1 \leq \ldots \leq \lambda_N.$$

Then

$$\lambda_1 \leq R(x) \leq \lambda_N \text{ for all } x \in \mathbb{R}^N.$$

Proof We apply *the Lagrange Multiplier Rule* to find extrema of the function

$$f(x) = \langle x, Ax \rangle$$

subject to

$$h(x) = \|x\|^2 - 1 = 0.$$

By a direct calculation we see that

$$\lambda_1 \leq \langle x, Ax \rangle \leq \lambda_N \text{ for } \|x\|^2 = 1.$$

Hence the assertion follows.

▶ **Remark 1.4** Apart from equality constraints, one also considers inequality constraints in which case *the Lagrange Multiplier Rule* is known as *the Karush–Kuhn–Tucker Theorem*, see [5,41] for a detailed treatment of this subject with many examples, comments, and algorithms.

Some Basics from Functional Analysis and Function Spaces

Now we introduce the function spaces used in what follows. Since we assume some familiarity with functional analysis (we refer to [7,8,28,46]), mathematical analysis [11,12,45], and Lebesgue integration [48], the results in this chapter are presented in some short manner.

2.1 On the Convexity with Some Revision from the Calculus

Let E be a real Banach space. In the space E an interval with ends $x_1, x_2 \in E$ is defined as expected

$$[x_1, x_2] = \{x \in E : x = \alpha x_1 + (1 - \alpha)x_2, \ \alpha \in [0, 1]\}.$$

Set $C \subset E$ is called convex if each two distinct points in C can be connected by an interval contained in C.

The functional $F : E \to \mathbb{R} \cup \{+\infty\}$ is called **convex** provided that

$$\forall_{\lambda \in [0,1]} \forall_{x_1, x_2 \in E} \quad F(\lambda x_1 + (1 - \lambda)x_2) \leq \lambda F(x_1) + (1 - \lambda)F(x_2).$$

The functional $F : E \to \mathbb{R} \cup \{+\infty\}$ is called **strictly convex** provided that

$$\forall_{\lambda \in (0,1)} \forall_{x_1, x_2 \in E, x_1 \neq x_2} \quad F(\lambda x_1 + (1 - \lambda)x_2) < \lambda F(x_1) + (1 - \lambda)F(x_2).$$

We can restrict the definition of a convex function to any convex subset of E. Moreover, the restriction of a convex function to some convex subset of its domain remains convex.

M. Galewski, *Basics of Nonlinear Optimization*, Compact Textbooks in Mathematics, https://doi.org/10.1007/978-3-031-77160-6_2

Exercise 2.1

Assume that the functional $F : E \to \mathbb{R}$ is convex. Prove that sets F^α are also convex for all $\alpha \in \mathbb{R}$. Demonstrate that for a non-convex functional $F(x) = \|x\|^3$ the Lebesgue sets are convex as well.

We recall a bit about convexity on an Euclidean space from calculus courses mainly through remarks, exercises, and examples. By $I_1, I_2 \subset \mathbb{R}$ we mean any interval (closed or open):

▶ **Remark 2.1** Assume that $F : I_1 \times I_2 \to \mathbb{R}$ is convex. Then obviously (and to no surprise)

(a) for each fixed $y_0 \in I_2$ the function $x \mapsto F(x, y_0)$ is convex on I_1;
(b) for each fixed $x_0 \in I_1$ the function $y \mapsto F(x_0, y)$ is convex on I_2.

Note that if $F : I_1 \times I_2 \to \mathbb{R}$ is a function for which the aforementioned conditions hold, it need not be convex. Again this is to no surprise and is the case for a non-convex function

$$F(x, y) = x^4 - xy + \frac{3}{2}y^2.$$

The convexity is checked with the second-order test saying that if the Hesse matrix of a C^2 function of several variables is positive definite then it strictly convex, and it is convex if and only if the Hesse matrix is positive semidefinite.

Exercise 2.2

Assume that $I_1, I_2 \subset \mathbb{R}$ are intervals and functions $F_1 : I_1 \to \mathbb{R}$, $F_2 : I_2 \to \mathbb{R}$ are convex. Show that function $F : I_1 \times I_2 \to \mathbb{R}$

$$F(x, y) = F_1(x) + F_2(y)$$

is convex as well. Function F is strictly convex provided any of F_1, F_2 shares this property. Prove that $F(x, y) = x^4 - \ln y$ is strictly convex over $\mathbb{R} \times (0, +\infty)$.

Exercise 2.3

Consider the function $F : \mathbb{R}^2 \to \mathbb{R}$ defined by

$$F(x, y) = e^{x-y} + e^{y-x} + e^{x^2}.$$

Show that F is strictly convex and $(0, 0)$ is a global minimizer with $F(0, 0) = 3$. Is F a coercive function?

Example 2.1

Consider a C^2 function $f : \mathbb{R}^2 \to \mathbb{R}$ defined by

$$F(x, y) = e^{x-y} + e^{y-x}.$$

We compute that

$$\nabla F(x, y) = \left[e^{x-y} - e^{y-x}, e^{y-x} - e^{x-y} \right],$$

$$HF(x, y) = \begin{bmatrix} e^{x-y} + e^{y-x} & -e^{x-y} - e^{y-x} \\ -e^{x-y} - e^{y-x} & e^{x-y} + e^{y-x} \end{bmatrix}.$$

Equating the gradient to zero, we see that

$$C = \{(x, y) : x = y\}.$$

is the set of critical points. Since $\det HF(x, y) = 0$ we need to employ the eigenvalue test to check the convexity. By a direct calculation we find both eigenvalues: $2e^{x-y} + 2e^{y-x}$, 0. This implies that F is convex and therefore C consists of minimizers. Observe that for $(x, y) \in C$, we have $F(x, y) = 2$. Note that F is not coercive.

Exercise 2.4

Show that a C^2 function $F : \mathbb{R}^2 \to \mathbb{R}$ defined by

$$F(x, y) = e^{x-y} + e^{y+x}.$$

is strictly convex and without minimizers.

Exercise 2.5

Show that function $F(x) = -e^{-x^2-y^2}$ has a global minimizer at $(0, 0)$ but it is not convex.

Exercise 2.6

Find values of parameters $a, b, c, p \subset \mathbb{R}$ for which the following functions are convex:

(a) $F_1(x) = ax^2 + bx + c$, $F_1 : \mathbb{R} \to \mathbb{R}$;
(b) $F_2(x) = ae^{2x} + be^x + c$, $F_2 : \mathbb{R} \to \mathbb{R}$;
(c) $F_3(x, y) = \left(|x|^p + |y|^p \right)^{1/p}$, $F_3 : \mathbb{R} \times \mathbb{R} \to \mathbb{R}$;
(d) $F_4(x, y) = ax^2 + 2bxy + cy^2$, $F_4 : \mathbb{R} \times \mathbb{R} \to \mathbb{R}$.

After the aforementioned revision, we are able to proceed with considering convexity in infinite-dimensional spaces. The relation between convexity of a function and convexity of its epigraph is formulated as follows:

Theorem 2.1
Let E be a real Banach space. Functional $F : E \to \mathbb{R} \cup \{+\infty\}$ is convex if and only if the set Epi(F) *is convex.*

Proof Let $(x, \alpha_x), (y, \alpha_y) \in$ Epi (F) and let $\lambda \in [0, 1]$. We see that

$$\lambda\alpha_x + (1 - \lambda)\alpha_y \geq \lambda F(x) + (1 - \lambda)F(y) \geq F(\lambda x + (1 - \lambda)y).$$

Hence Epi (F) is convex.

Now let Epi (F) be convex and let $(x, F(x)), (y, F(y)) \in$ Epi (F). Then for $\lambda \in [0, 1]$

$$(\lambda x + (1 - \lambda)y, \lambda F(x) + (1 - \lambda)F(y)) \in \text{Epi } (F)$$

which implies that

$$\lambda F(x) + (1 - \lambda)F(y) \geq F(\lambda x + (1 - \lambda)y).$$

Although convexity is a geometric notion it has a great impact on continuity. We recall that if $F : (a, b) \to \mathbb{R}$ is convex, then it is absolutely continuous. We have the following:

Theorem 2.2   (Continuity of a Convex Functional)
Let E be a real Banach space. Let $C \subset E$ be open and convex and let $F : C \to \mathbb{R}$ be convex and bounded from above in some neighborhood of $x_0 \in C$. Then F is continuous at x_0.

Proof We can always assume that $F(0) = 0$ and reduce the proof to the case of checking continuity at $x_0 = 0$. We take an open ball $B(0, r)$ centered at 0 with radius r. Assume that there is a constant $M > 0$ such that

$$F(x) \leq M \text{ for all } x \in B(0, r). \tag{2.1}$$

Fix $\varepsilon \in (0, 1)$ and take any $x \in B(0, \varepsilon r)$. Observe that

$$\frac{x}{\varepsilon} \in B(0, r) \quad \text{and} \quad -\frac{x}{\varepsilon} \in B(0, r).$$

It follows by the convexity of F that

$$F(x) = F\left((1 - \varepsilon)0 + \varepsilon\frac{x}{\varepsilon}\right) \le (1 - \varepsilon)F(0) + \varepsilon F\left(\frac{x}{\varepsilon}\right) = \varepsilon F\left(\frac{x}{\varepsilon}\right) \le \varepsilon M$$

and also

$$0 = (1 + \varepsilon) F(0) = (1 + \varepsilon) F\left(\frac{1}{1 + \varepsilon}x + \frac{\varepsilon}{1 + \varepsilon}\left(-\frac{x}{\varepsilon}\right)\right) \le F(x) + \varepsilon F\left(-\frac{x}{\varepsilon}\right).$$

Summarizing for any $x \in B(0, \varepsilon r)$ we obtain the following estimation

$$|F(x)| \le \varepsilon M$$

which implies the continuity of F at $x_0 = 0$.

▶ **Remark 2.2** From the proof of Theorem 2.2, we can further learn that convex functional bounded from above on some neighborhood is in fact locally Lipschitz at least on this neighborhood, see Proposition 2.2.6 from [14]. We will show that F is Lipschitz around 0. Indeed, with notation of the above proof, let us take some $x_1, x_2 \in B(0, r/2)$. Put $\alpha = \|x_1 - x_2\|$ and define

$$x_3 = x_2 + \frac{r}{2\alpha}(x_2 - x_1) \in B(0, r).$$

We find that

$$x_2 = \frac{r}{r + 2\alpha}x_1 + \frac{2\alpha}{r + 2\alpha}x_3$$

and by the convexity

$$F(x_2) \le \frac{r}{r + 2\alpha}F(x_1) + \frac{2\alpha}{r + 2\alpha}F(x_3).$$

Next using (2.1) we have

$$F(x_2) - F(x_1) \le \frac{2\alpha}{r + 2\alpha}(F(x_3) - F(x_1)) \le \frac{2\alpha}{r}|F(x_3) - F(x_1)| \le \frac{4M}{r}\alpha.$$

Since we can reverse the order of x_1 and x_2 in the above, we get using the definition of α

$$|F(x_2) - F(x_1)| \leq \frac{4M}{r} \|x_1 - x_2\|.$$

A direct corollary now follows:

Corollary 2.1

Let E be a real Banach space. Let $F : E \to \mathbb{R} \cup \{+\infty\}$ be lower semicontinuous and convex. Then it is locally Lipschitz continuous over the interior of its effective domain.

The following exercises are to be treated as a word of caution when dealing with convex functions that take $+\infty$.

Exercise 2.7

Consider function $F : \mathbb{R}^2 \to \mathbb{R} \cup \{+\infty\}$ defined by

$$F(x, y) = \begin{cases} \frac{x^2}{y}, & y > 0, \\ 0, & (x, y) = 0, \\ +\infty, & \text{otherwise.} \end{cases} \tag{2.2}$$

Show that this function is convex and its epigraph is closed (i.e., it is lower semicontinuous.)

Exercise 2.8

Show that function F given by (2.2) is not continuous at $(0, 0)$, i.e., F is continuous only in the interior of its effective domain.

2.2   On the Weak Convergence

We start with a definition of a weak convergence and some of its related properties. This simplifies the setting but it is what is required for the scope of this book. For the general case, see [46].

Definition 2.1 (Weak Convergence)

Let E be a real Banach space. A sequence $(x_n) \subset E$ is said to be *weakly convergent* to an element $x_0 \in E$ if

$$\langle f, x_n \rangle \to \langle f, x_0 \rangle \text{ as } n \to +\infty.$$

for each $f \in E^*$. We will write

$$x_n \rightharpoonup x_0 \text{ in } E$$

and we will call element x_0 a weak limit of a sequence (x_n).

The limit in the sense of a norm will sometimes be addressed as a strong limit. In this case we will write $x_n \to x_0$ in E which means $\lim_{n \to +\infty} \|x_n - x_0\| = 0$. The weak limit is uniquely defined. Indeed, if $x_n \rightharpoonup x_0$ and $x_n \rightharpoonup y_0$, then for any $f \in E^*$ we have

$$0 = \lim_{n \to +\infty} \langle f, x_n - x_n \rangle = \lim_{n \to +\infty} \langle f, x_n \rangle - \lim_{n \to +\infty} \langle f, x_n \rangle = \langle f, x_0 - y_0 \rangle .$$

This implies that $x_0 = y_0$. Moreover, a weakly convergent sequence is necessarily (norm) bounded, see Theorem 16.14 c) from [28].

Exercise 2.9

Show that in a finite-dimensional space the weak and the strong convergence are equivalent.

Exercise 2.10

Show that if $x_n \to x_0$ in E, then $x_n \rightharpoonup x_0$ in E.

Exercise 2.11

Show that the weak limit obeys the same arithmetic laws as the strong one.

Exercise 2.12

Show that if $a_n \to a_0$ in $\mathbb{R}$ and $x_n \rightharpoonup x_0$ in E, then $a_n x_n \rightharpoonup a_0 x_0$ in E.

Exercise 2.13

Let E be a reflexive space. Assume that $(x_n) \subset E$ and $(f_n) \subset E^*$. Show that if either of these sequences converges weakly and the other one strongly to their respective limits $x_0 \in E$ and $f_0 \in E^*$, then

$$\lim_{n \to +\infty} \langle f_n, x_n \rangle = \langle f_0, x_0 \rangle .$$

Note that in the aforementioned exercise the assumption that at least one of the above sequences converges strongly, is crucial as well as the assumption that E is reflexive.

Exercise 2.14

Let E be a real Hilbert space. Let $(x_n) \subset E$ and let $x_0 \in E$. Show (using the properties of the scalar product) that if $x_n \rightharpoonup x_0$ and if $\|x_n\| \to \|x_0\|$, then $x_n \to x_0$.

Definition 2.2

Let E be a real Banach space. We say that a subset of E is sequentially weakly closed if it contains limits of its all weakly convergent sequences.

Exercise 2.15

Show that any sequentially weakly closed subset of E is (norm) closed.

The reverse of the aforementioned exercise is not true unless some additional conditions are imposed on the set in question. Typically the following result is used for determining when closed sets are also sequentially weakly closed (see Theorem 3.7 and Corollary 3.8 in [8]):

Lemma 2.1
Let E be a real Banach space. Let $D \subset E$ be closed and convex. Then D is sequentially weakly closed.

By the Riesz Representation Theorem $\left(l^2\right)^*$ is identified with l^2. We follow with a standard example of a closed set, which is not weakly closed (the unit sphere):

Example 2.2

Let us take a sequence (e_n) from a unit sphere S_{l^2} of l^2 given as follows

$$e_1 = (1, 0, 0, \ldots), \ e_2 = (0, 1, 0, 0, ..), \ldots$$

This sequence is bounded and it does not contain any strongly convergent subsequence. On the other hand, by a direct calculation using the necessary convergence condition, we demonstrate that (e_n) is weakly convergent to 0_{l^2}.

Exercise 2.16

Consider the set $S_{l^2} \cup \{0_{l^2}\}$. Is it sequentially weakly closed? Determine the sequential weak closure of S_{l^2}.

Now we proceed to the counterpart of the lower semicontinuity in the presence of the weak convergence. The results for checking the sequential weak lower semicontinuity will be of use later on when we will proceed to the version of the Weierstrass Theorem in infinite dimensional spaces. Since the proofs mostly resemble their finite-dimensional counterparts, we do not give them but rather leave as exercises:

Definition 2.3 (Sequential Weak Lower Semicontinuity)

Let E be a real Banach space. We say that a functional $F : E \to \mathbb{R}$ is sequentially weakly lower semicontinuous if

$$\liminf_{n \to +\infty} F(x_n) \geq F(x_0)$$

for any $x_0 \in E$ and any sequence $(x_n) \subset E$ such that $x_n \rightharpoonup x_0$.

Exercise 2.17

Let E be a real Banach space. Prove that functional $F : E \to \mathbb{R}$ is sequentially weakly lower semicontinuous if and only if sets F^α are sequentially weakly closed for $\alpha \in \mathbb{R}$.

Exercise 2.18

Prove that a norm in any real Banach space is sequentially weakly lower semicontinuous.

Definition 2.4

Let E be a real Banach space. We say that a functional $F : E \to \mathbb{R}$ is sequentially weakly continuous if

$$\lim_{n \to +\infty} F(x_n) = F(x_0)$$

for any $x_0 \in E$ and any sequence $(x_n) \subset E$ such that $x_n \rightharpoonup x_0$.

Exercise 2.19

Let $F : E \to \mathbb{R}$ be a sequentially weakly continuous functional. Prove that F is continuous.

Example 2.3 (Schur Property)

We say that E has **Schur's property** if $x_n \rightharpoonup x_0$ implies that $x_n \to x_0$. The space l^1 consisting of absolutely convergent numerical series has Schur's property, i.e., weakly and strongly convergent sequences coincide. A norm in a space without the Schur property serves as an example of a functional, which is continuous but which is not sequentially weakly continuous. Indeed, assuming that the norm is weakly continuous at 0, we would have that

$$\lim_{n \to +\infty} \|x_n\| = 0$$

for any sequence $(x_n) \subset E$ with $x_n \rightharpoonup 0$. But is a contradiction.

Now we concentrate on another sufficient conditions for a functional to be sequentially weakly lower semicontinuous.

> **Theorem 2.3 (Sufficient Condition for Weak Lower Semicontinuity)**
> *Let E be a real Banach space. Assume that functional $F : E \to \mathbb{R}$ is lower semicontinuous and convex. Then F is sequentially weakly lower semicontinuous.*

Proof Since F is lower semicontinuous, it follows that sets F^α are closed for all $\alpha \in \mathbb{R}$. Since F is convex, it follows that sets F^α are also convex for all $\alpha \in \mathbb{R}$. In view of Lemma 2.1, we see that since all F^α are closed and convex, these are sequentially weakly closed. But this implies that functional F is sequentially weakly lower semicontinuous.

Exercise 2.20

Prove that functional F is sequentially weakly lower semicontinuous iff set $\mathrm{Epi}(F)$ is sequentially weakly closed.

Further on we will need some information about function spaces. The space $L^1(0, 1)$ consists of Lebesgue integrable functions defined on $[0, 1]$ with values in $\mathbb{R}$ which when endowed with the norm

$$\|u\|_{L^1} = \int_0^1 |u(t)| \, dt$$

becomes a Banach space (although not reflexive). By $L^2(0, 1)$, denoted also L^2, we understand the space of all Lebesgue measurable functions defined on $[0, 1]$ with values in $\mathbb{R}$ which are square integrable. When endowed with a scalar product

$$(u, v)_{L^2} = \int_0^1 u(t) v(t) \, dt \text{ for } u, v \in L^2(0, 1),$$

it becomes a separable Hilbert space. A well-known Schwarz Inequality (valid in any Hilbert space) is as follows:

$$(u, v)_{L^2} \le \|u\|_{L^2} \|v\|_{L^2}, \text{ for } u, v \in L^2(0, 1),$$

where

$$\|u\|_{L^2} = \sqrt{\int_0^1 |u(t)|^2 \, dt} \text{ for } u \in L^2(0, 1).$$

The space $\left(L^2 (0, 1)\right)^*$ is identified with $L^2 (0, 1)$ via the Riesz Representation Theorem. Some properties of convergent (both strongly and weakly) sequences from $L^2 (0, 1)$ now follow given as lemmas and exercises.

Exercise 2.21

Define $(u_n) \subset L^2 (0, 1)$ by

$$u_n (t) = \sqrt{n} \chi_{\left(0, \frac{1}{n}\right)},$$

where $\chi_A : A \to \mathbb{R}$ denotes the indicator function of set $A \subset E$, i.e.

$$\chi_A (t) = \begin{cases} 1, \; t \in A, \\ 0, \; t \notin A. \end{cases}$$

Prove that $u_n \rightharpoonup 0$ and also $u_n (t) \to 0$ for a.e. $t \in [0, 1]$.

The aforementioned exercise should be compared with the following:

Lemma 2.2

Take a sequence $(u_n) \subset L^2 (0, 1)$ which is weakly convergent to some u_0, that is

$$\lim_{n \to +\infty} \int_0^1 u_n (t) v (t) \, dt = \int_0^1 u_0 (t) v (t) \, dt \; \text{for all } v \in L^2 (0, 1).$$

Assume also that (u_n) converges a.e. on $[0, 1]$ to some $\tilde{u}$, that is

$$\lim_{n \to +\infty} u_n (t) = \tilde{u} (t) \; \text{for a.e. } t \in [0, 1]. \tag{2.3}$$

Then $\tilde{u} (t) = u_0 (t)$ for a.e. $t \in [0, 1]$.

Proof Indeed, since (u_n) converges weakly it is bounded, i.e., there is a constant $M > 0$ such that

$$\int_0^1 u_n^2 (t) \, dt \leq M \text{ for all } n \in \mathbb{N}.$$

Using the Fatou Lemma, we obtain what follows

$\int_0^1 \tilde{u}^2 (t) \, dt = \int_0^1 (\lim_{n \to +\infty} u_n (t))^2 \, dt =$

$\int_0^1 (\liminf_{n \to +\infty} u_n (t))^2 \, dt \leq \liminf_{n \to +\infty} \int_0^1 (u_n (t))^2 \, dt \leq M$

which shows that $\tilde{u} \in L^2(0, 1)$. Now redefining the sequence (u_n) on the null set we can assume that the convergence in (2.3) holds everywhere on $[0, 1]$. We define the following measurable sets

$$E_k = \left\{ t \in [0, 1] : \sup_{n \geq k} |u_n(t)| \geq k \right\}, \quad k = 1, 2, \ldots$$

Then we see that

$$\bigcap_k E_k = \left\{ t \in [0, 1] : \limsup_{n \to +\infty} |u_n(t)| = +\infty \right\}.$$

is the null set, i.e.

$$\text{mes} \left(\bigcap_k E_k \right) = 0.$$

Using the obvious inclusion $E_k \supset E_{k+1}, k = 1, 2, \ldots$, we get

$$\lim_{k \to +\infty} \text{mes}(E_k) = 0. \tag{2.4}$$

Let now $t \in [0, 1] \setminus E_k$. Then we have for any $w \in L^2(0, 1)$

$$|u_n(t)w(t)| \leq k|w(t)|, \quad \text{for all } n \geq k.$$

Since the sequence (u_n) converges weakly to u_0 and a.e. to $\tilde{u}$ we get by the Lebesgue Dominated Convergence Theorem

$$\lim_{n \to +\infty} \int_{[0,1]\setminus E_k} u_n(t)w(t)dt = \int_{[0,1]\setminus E_k} u_0(t)w(t)dt = \int_{[0,1]\setminus E_k} \tilde{u}(t)w(t)dt.$$

This means that $\tilde{u}(t) = u_0(t)$ for $t \in [0, 1] \setminus E_k$. Relation (2.4) now implies the assertion.

The aforementioned result should be further commented as follows:

▶ **Remark 2.3** If $(u_n) \subset L^2(0, 1)$ is norm convergent to $u_0 \in L^2(0, 1)$, then it contains a subsequence, which is convergent a.e. on $[0, 1]$, again to the same function u_0. If $(u_n) \subset L^2(0, 1)$ is weakly convergent to some $u_0 \in L^2(0, 1)$, it does not mean that it contains any subsequence converging almost everywhere.

Exercise 2.22

Define $(u_n) \subset L^2(0, 1)$ by

$$u_n(t) = \sin(2n\pi t).$$

Prove that (u_n) is nowhere convergent on $(0, 1)$ but that $u_n \rightharpoonup 0$. Hint: Recall the Riemann–Lebesgue Lemma from the classical theory of Fourier expansions, see [12]).

An example of a non-reflexive space is $L^\infty(0, 1)$. We say that a measurable function $u : [0, 1] \to \mathbb{R}$ belongs to $L^\infty(0, 1)$ if there exists a constant $c > 0$ such that

$$|u(t)| \le c \text{ for a.e. } t \in [0, 1].$$

L^∞ becomes a Banach spaces when endowed with a following norm

$$\|u\|_{L^\infty} = \inf\{c : |u(t)| \le c \text{ for a.e. } t \in [0, 1]\}.$$

If $\Omega \subset \mathbb{R}^N$ is a measurable set we see that it is immediate to define the spaces $L^i(\Omega)$, $L^\infty(\Omega)$, for $i = 1, 2$. The Reader is invited to work the details up or check in the literature suggested, see [7, 8, 20, 28, 46, 48, 52].

2.3 Niemytskij Operator and the Krasnosel'skii-Type Theorem

The Krasnosel'skii-type theorem will be used to prove the continuity of nonlinear superposition operators, which concern the so-called Niemytskij operator.

Definition 2.4 (L^p-Carathéodory Function)

Let $p \ge 1$. We say that $f : [0, 1] \times \mathbb{R} \to \mathbb{R}$ is an L^p-*Carathéodory* function if the following conditions are satisfied:

(a) $t \mapsto f(t, x)$ is measurable on $[0, 1]$ for each fixed $x \in \mathbb{R}$,
(b) $x \mapsto f(t, x)$ is continuous on $\mathbb{R}$ for a.e. $t \in [0, 1]$,
(c) for each $d \in \mathbb{R}^+$ function $t \mapsto \max_{|x| \le d} |f(t, x)|$ belongs to $L^p(0, 1)$.

When f satisfies only first two conditions of the aforementioned definition, it is called a Carathéodory function.

Exercise 2.23

Assume that $f : [0, 1] \times \mathbb{R} \to \mathbb{R}$ is componentwise continuos. Check if it is a Carathéodory function.

Exercise 2.24

Show that the following are L^p−Carathéodory functions $f : [0, 1] \times \mathbb{R} \to \mathbb{R}$:

(a) $f(t, x) = f_1(t) \cdot \arctan x$, $f_1 \in L^p (0, 1)$;
(b) $f(t, x) = f_1(t)x^4 + f_2 (t) x^3 + \arctan(t \cdot x)$, f_1, $f_2 \in f \in L^p (0, 1)$.

Let $\mathcal{M}$ be the set of all measurable functions $u : [0, 1] \to \mathbb{R}$. The composition of two measurable functions need not be measurable, but when $u \in \mathcal{M}$ and $f : [0, 1] \times \mathbb{R} \to \mathbb{R}$ is a Carathéodory function, it follows that $t \mapsto f (t, u (t))$ is measurable. The Niemytskij operator induced by f is an operator

$$N_f : \mathcal{M} \to \mathcal{M}$$

such that

$$N_f (u) (\cdot) = f (\cdot, u (\cdot)) \text{ a.e. on } [0, 1].$$

We now proceed to the Krasnosel'skii Theorem on the continuity of the Niemytskij operator, which provides both a necessary and sufficient condition. We will use and prove the sufficient condition only.

Theorem 2.4 (Krasnosel'skii Theorem)
Let $p_1, p_2 \geq 1$. Let $f : [0, 1] \times \mathbb{R} \to \mathbb{R}$ be a Carathéodory function for which there exist a constant $b \geq 0$ and a nonnegative function $a \in L^{p_2} (0, 1)$ satisfying

$$|f (t, x)| \leq a (t) + b |x|^{\frac{p_1}{p_2}} \text{ for a.e. } t \in [0, 1] \text{ and all } x \in \mathbb{R}. \qquad (2.5)$$

Then the Niemytskij operator $N_f : L^{p_1} (0, 1) \to L^{p_2} (0, 1)$ induced by f is well defined and continuous.

Proof Using the Minkowski Inequality (or the triangle inequality) we easily see that operator N_f is well defined by (2.5). We take any sequence $u_n \to \bar{u}$ in $L^{p_1} (0, 1)$. There exists a subsequence that converges almost everywhere to $\bar{u}$ and which has a further subsequence that $f (t, u_n (t)) \to f (t, \bar{u} (t))$ for a.e. $t \in [0, 1]$. Note that we do not renumber the subsequences. Moreover, there is a function $h \in L^{p_1} (0, 1)$ such that $|u_n (t)| \leq h (t)$ for *a.e.* $t \in [0, 1]$, again up to a subsequence. Then (2.5) implies that

$$|f (t, u_n (t))| \leq a (t) + b |h (t)|^{\frac{p_1}{p_2}} \text{ for a.e. } t \in [0, 1].$$

Hence we obtain the assertion by the Lebesgue Dominated Convergence Theorem.

▶ **Remark 2.4** Note that in the aforementioned considerations, we can replace $[0, 1]$ with a measurable set $\Omega \subset \mathbb{R}^N$. Then the assertion of *the Krasnosel'skii Theorem* says that $N_f : L^{p_1}(\Omega) \to L^{p_2}(\Omega)$ induced by f is well defined and continuous. Moreover, we can replace constant b with some positively valued function $b \in L^{\infty}(0, 1)$.

Exercise 2.25

Prove the following version of the Krasnosel'skii Theorem: Let $p_1, p_2 \geq 1$. Assume that $f : [0, 1] \times \mathbb{R} \to \mathbb{R}$ is a Carathéodory function. If for any sequence $(u_n) \subset L^{p_1}(0, 1)$ convergent to $\overline{u} \in L^{p_1}(0, 1)$ there exists a function $h \in L^{p_2}(0, 1)$ such that

$$|f(t, u_n)| \leq h(t), \quad \text{for } n \in \mathbb{N} \text{ and } a.e. \ t \in [0, 1],$$

then the Niemytskij operator induced by f is well defined and continuous.

2.4   On the Space $C[0, 1]$

The space $C[0, 1]$ is defined as follows

$$C[0, 1] = \{u : u : [0, 1] \to \mathbb{R} \text{ is continuous}\}.$$

If $u, v \in C[0, 1]$ and $\lambda \in \mathbb{R}$, then the sum and the scalar multiple of functions is defined pointwise, i.e.,

$$(u + v)(t) = u(t) + v(t), \ (\lambda u)(t) := \lambda u(t) \text{ for } t \in [0, 1].$$

We know from elementary analysis that $f + g$, $\lambda v \in C[0, 1]$ again which allows us to demonstrate that $C[0, 1]$ is a linear space over $\mathbb{R}$. When considered with the norm

$$\|u\|_{\infty} = \sup_{t \in [0,1]} |u(t)| \text{ for } u \in C[0, 1]$$

it becomes a Banach space (for the proof see Lemma 5.1 later on), although it is non-relfexive, and it is not easy to describe its dual as well. When we mention a norm in $C[0, 1]$ we mean just the supremum norm if it is not said otherwise. The convergence with respect to the supremum norm is called the uniform convergence and is denoted sometimes by symbol $\rightrightarrows$. The uniform convergence implies the pointwise convergence, i.e., it implies that

$$\lim_{n \to +\infty} u_n(t) = u_0(t) \text{ for all } t \in [0, 1]$$

but not the vice versa as is known from the example of a sequence $u_n(t) = t^n$ on $[0, 1]$. It is immediate to shift the uniform convergence from $[0, 1]$ to any set $A \subset \mathbb{R}$ by saying that the sequence of functions (u_n) converges uniformly to u on A provided that

$$\forall_{\varepsilon > 0} \exists_{N := N(\varepsilon)} \forall_{t \in A} \forall_{n \geq N} |u_n(t) - u(t)| < \varepsilon.$$

We say that (u_n) converges almost uniformly to u on $A \subset \mathbb{R}$ if $u_n \rightrightarrows u$ on each bounded and closed subset of A. The following exercises serve as some practice concerning the convergence in $C[0, 1]$.

Exercise 2.26

Study uniform and pointwise convergence for the sequence of maps:

$$u_n(t) = nt \left(1 - t^2\right)^n, \quad t \in [0, 1].$$

Does the following formula hold?

$$\lim_{n \to +\infty} \int_0^1 u_n(t) \mathrm{d}t = \int_0^1 \lim_{n \to +\infty} u_n(t) \mathrm{d}t.$$

Exercise 2.27

Show that the sequence

$$u_n(t) = \frac{nt}{1 + n^2 t^2}$$

converges almost uniformly on $(0, 1]$, but not on $[0, 1]$. Study the case of the sequence

$$u_n(t) = \frac{t}{1 + n^2 t^2}$$

on the same sets. Note that both converge pointwise to the same function.

We can equip $C[0, 1]$ with the following norms

$$\|u\|_{L^1} = \int_0^1 |u(t)| \, \mathrm{d}t$$

and

$$\|u\|_{L^2} = \int_0^1 |u(t)|^2 \, \mathrm{d}t$$

both making $C[0, 1]$ into a normed space which is however not complete. This is seen by considering the sequence $u_n(t) = t^n$ whose pointwise limit is not continuous on $[0, 1]$.

Exercise 2.28

Prove that

$$\|u\|_{L^1} \leq \|u\|_{L^2} \leq \|u\|_\infty \text{ for any } u \in C[0, 1].$$

Example 2.4

Now let $f \in C([0, 1] \times \mathbb{R})$. We consider the following integral functional $F : C[0, 1] \to \mathbb{R}$

$$F(u) = \int_0^1 f(t, u(t)) \, dt$$

which is obviously well defined (i.e. finite) for all $u \in C[0, 1]$. We will establish its continuity at any $u_0 \in C[0, 1]$ with respect to the norm $\|\cdot\|_\infty$. Let us fix $\varepsilon > 0$. We will show that there is a $\delta \in (0, 1)$ such that

$$\|u - u_0\|_\infty < \delta \Rightarrow \int_0^1 |f(t, u(t)) - f(t, u_0(t))| \, dt < \varepsilon.$$

Since u_0 is continuous on $[0, 1]$ by *the Weierstrass Theorem* there is some $c_1 > 0$ that $|u_0(t)| \leq c_1$ for all $t \in [0, 1]$. Then there is some $c > 0$ such that for all $u \in C[0, 1]$ with $\|u - u_0\|_\infty \leq 1$ it holds that

$$\max\{|u_0(t)|, |u(t)|\} \leq c \text{ for all } t \in [0, 1].$$

Since function f is uniformly continuous on $[0, 1] \times [-c, c]$ (it is continuous and so the Cantor Theorem applies), there is some $\delta \in (0, 1)$ such that

$$|f(t, z) - f(t, z_0)| < \varepsilon$$

for any pairs $(t, z_0), (t, z)$ from $[0, 1] \times [-c, c]$ with $|z - z_0| < \delta$. Then taking $u \in C[0, 1]$ such that $\|u - u_0\|_\infty < \delta$ one has

$$|F(u) - F(u_0)| \leq \int_0^1 |f(t, u(t)) - f(t, u_0(t))| \, dt < \varepsilon.$$

The aforementioned example covers a large class of functionals and says that the (joint) continuity of the integrand implies the continuity of the integral functional considered on C[0, 1].

Exercise 2.29

Let $t_0 \in [0, 1]$ be fixed. Consider *the evaluation functional* $F : C[0, 1] \to \mathbb{R}$ defined by

$$F(u) = u(t_0).$$

Show that F is continuous.

Exercise 2.30

Consider C[0, 1] with $\|\cdot\|_{L^1}$ −norm. Check if *the evaluation functional* $F(u) = u(0)$ is continuous.

Exercise 2.31

Establish the continuity of the functional $F : C[0, 1] \to \mathbb{R}$ defined by

$$F(u) = \int_0^1 (\sin t)\, u(t)\, dt$$

when C[0, 1] is considered with the following norms: $\|\cdot\|_{L^i}$ for $i = 1, 2$ and $\|\cdot\|_\infty$.

Exercise 2.32

Let $E = (C[0, 1])^d$ be the space of d-dimensional vector functions with components in C[0, 1]. Prove that it is a normed space with respect to the following norms

$$\|u\|_1 = \sup_{t\in[0,1]} |u(t)|, \quad \|u\|_2 = \sum_{j=1}^d \sup_{t\in[0,1]} |u_j(t)|, \quad \|u\|_3 = \sum_{j=1}^d \int_0^1 |u_j(t)|\, dt,$$

and finally

$$\|u\|_4 = \sqrt{\int_0^1 \sum_{j=1}^d |u_j(t)|^2\, dt}.$$

Which of the aforementioned norms makes E into a Banach space.

We will also need the space $C^1[0, 1]$ consisting of functions defined and continuously differentiable on $[0, 1]$, while at 0 and 1 we understand, as usual, one-sided derivatives. Such a space is conveniently normed by

$$\|u\|_M = \sup_{t \in [0,1]} \left(|u(t)| + |u'(t)| \right) \tag{2.6}$$

or

$$\|u\|_1 = \sup_{t \in [0,1]} |u'(t)| + |u(0)| \tag{2.7}$$

both making it into a Banach space. The convergence with respect to either norm is the uniform convergence of both functions and their derivatives over $[0, 1]$.

Exercise 2.33

Prove that the convergence in $\| \cdot \|_1$ defined by (2.7) implies the uniform convergence of both functions and their derivatives over $[0, 1]$.

Exercise 2.34

Let $f \in C\left([0, 1] \times \mathbb{R} \times \mathbb{R} \right)$ and consider the following integral functional $F :$ $C^1[0, 1] \to \mathbb{R}$

$$F(u) = \int_0^1 f\left(t, u(t), u'(t) \right) dt$$

Prove that it is continuous with respect to the norm given by (2.6).

Exercise 2.35

Prove that formula $\|u\|_2 = \sup_{t \in [0,1]} |u'(t)|$ does not define a norm on $C^1[0, 1]$. Can we norm $C^1[0, 1]$ with the norm $\| \cdot \|_\infty$ inherited from $C[0, 1]$. Will it still be a Banach space?

Exercise 2.36

Consider the space $\left(C^1[0, 1] \right)^d$ of d-dimensional vector functions with components in $C^1[0, 1]$. Prove that it can be normed by

$$\|u\|_M = \sup_{t \in [0,1]} \left(|u(t)| + |u'(t)| \right), \text{ where } |u(t)| = \sqrt{\sum_{j=1}^d |u_j(t)|^2}$$

and also by

$$\|u\|_1 = \sum_{j=1}^{d} \sup_{t \in [0,1]} \left(\left| u_j(t) \right| + \left| u'_j(t) \right| \right).$$

By $C_0^1 [0, 1]$ we understand the following space

$$C_0^1 [0, 1] = \left\{ u \in C^1 [0, 1] : u(0) = u(1) = 0 \right\}. \tag{2.8}$$

Exercise 2.37

Prove that $C_0^1 [0, 1]$ is a closed subspace of $C^1 [0, 1]$.

▶ **Remark 2.5** We should pay attention to the difference between $C[0, 1]$ and $C(0, 1)$ which is the space of continuos functions defined on $(0, 1)$. It is obvious that $C(0, 1)$ is a linear space but it has no obvious choice for a norm. We note that for the function $u(t) = \frac{1}{t}$ both norms $\|\cdot\|_\infty$ and $\|\cdot\|_{L^i}$ with $i = 1, 2$ are infinite when considered over $(0, 1)$.

Exercise 2.38

Compose the function tan with $u(t) = at + b$ for suitable a, b in order to produce the function belonging to $C(0, 1)$ but not to $C[0, 1]$.

2.5   On Absolutely Continuous Functions

There are two major approaches while introducing the space $H_0^1(0, 1)$. The one which we now adopt is via absolutely continuous functions and follows from [37], while the second one is based on the notion of weak differentiability and is described in [8] in detail and also sketched in Sect. 5.1.

Definition 2.6 (Absolutely Continuous Function)
A function $u : [0, 1] \to \mathbb{R}$ is said to be *absolutely continuous on* $[0, 1]$ if, given $\varepsilon > 0$, there exists some $\delta > 0$ such that

$$\sum_{i=1}^{n} |u(b_i) - u(a_i)| < \varepsilon$$

whenever $\{[a_i, b_i] : i = 1, \ldots, n\}$ is a finite collection of mutually disjoint subintervals of $[0, 1]$ with

$$\sum_{i=1}^{n} |b_i - a_i| < \delta.$$

Exercise 2.39

Prove that an absolutely continuous function on $[0, 1]$ is uniformly continuous.

Exercise 2.40

Prove that a Lipschitz continuous function on $[0, 1]$ as well as a $C^1[0, 1]$ function is absolutely continuous there.

Exercise 2.41

Show that function $u(t) = \sqrt{t}$ does not satisfy the Lipschitz condition on $[0, 1]$.

Now we proceed to an example of an absolutely continuos function, which is not locally Lipschitz.

Example 2.5

We will show that function $u(t) = \sqrt{t}$ is absolutely continuos on $[0, 1]$. Let $\varepsilon \in (0, 1)$ and put $\delta := \varepsilon^2/2$. Let $\{[a_i, b_i] : i \leqslant n\} \subset [0, 1]$ be such a family of closed and mutually disjoint subintervals of $[0, 1]$ that $\sum_{i=1}^{n} (b_i - a_i) < \delta$ and that the following monotonicity condition holds

$$a_i < b_i < a_{i+1} < b_{i+1} \text{ for } i \in \{1, \ldots, n-1\}.$$

We can assume that there is some $k \in \{1, \ldots, n-1\}$, for which $b_k \leqslant \varepsilon^2/4 \leqslant a_{k+1}$. (When $\varepsilon^2/4 \in (a_i, b_i)$, we divide (a_i, b_i) into subintervals $(a_i, \varepsilon^2/4)$ and $(\varepsilon^2/4, b_i)$.) Then

$$\sum_{i=1}^{n} \left(\sqrt{b_i} - \sqrt{a_i}\right) = \sum_{i=1}^{k} \left(\sqrt{b_i} - \sqrt{a_i}\right) + \sum_{i=k+1}^{n} \left(\sqrt{b_i} - \sqrt{a_i}\right)$$

$$\leqslant \sqrt{b_k} + \sum_{i=k+1}^{n} \frac{b_i - a_i}{2\sqrt{t_i}}$$

$$\leqslant \frac{\varepsilon}{2} + \frac{1}{2\sqrt{\varepsilon^2/4}} \sum_{i=k+1}^{n} (b_i - a_i) < \frac{\varepsilon}{2} + \frac{\delta}{\varepsilon} = \frac{\varepsilon}{2} + \frac{\varepsilon}{2} = \varepsilon,$$

where $t_i \in (a_i, b_i)$ for $i \in \{k+1, \ldots, n\}$ are points obtained by the application of *the Lagrange Mean Value Theorem* to function u over $[a_i, b_i]$ (we have $t_i \geqslant a_{k+1} \geqslant \varepsilon^2/4$, which implies that $1/\sqrt{t_i} \leqslant 1/\sqrt{\varepsilon^2/4}$).

There exist functions that are uniformly continuous and not absolutely continuous. There exists a continuous monotone function that is not absolutely continuous.

Exercise 2.42

Check that the function $u : [0, 1] \to \mathbb{R}$ defined by

$$u(t) = \begin{cases} t \sin\left(\frac{1}{t}\right), & t \in (0, 1], \\ 0, & t = 0 \end{cases}$$

is uniformly continuous and not absolutely continuous.

Exercise 2.43

Prove that *the Cantor function* is not absolutely continuous. See [15] for a survey paper about the Cantor function.

Exercise 2.44

Prove that any convex function $f : [0, 1] \to \mathbb{R}$ is absolutely continuous.

We have the following results about absolutely continuous functions, which we provide without proofs:

Lemma 2.3

Let u be an absolutely continuous function on $[0, 1]$. Then the derivative $\dot{u}(t)$ exists for almost every $t \in [0, 1]$. On the other hand, if function u is integrable on $[0, 1]$, then the function U defined by

$$U(t) := \int_0^t u(s)\mathrm{d}s, \quad 0 \le t \le 1$$

is absolutely continuous on $[0, 1]$.

Lemma 2.4

A function U on $[0, 1]$ is absolutely continuous if and only if

$$U(t) = U(0) + \int_0^t u(s)\mathrm{d}s \ \textit{for } t \in [0, 1]$$

for some integrable function u on $[0, 1]$.

With the aforementioned lemmas, we will usually write for an absolutely continuous function

$$u(t) = u(0) + \int_0^t \dot{u}(s)ds \text{ for } t \in [0, 1].$$

Proposition 2.1
Let u and v be two absolutely continuous functions defined on $[0, 1]$. Then $u + v$, $u - v$, and uv are absolutely continuous on $[0, 1]$. If, in addition, there exists a constant $C > 0$ such that $|v(t)| \geq C$ for all $t \in [0, 1]$, then u/v is absolutely continuous on $[0, 1]$. Moreover, the following integration by parts formula holds

$$\int_0^1 \dot{u}(t)v(t)dt = u(1)v(1) - u(0)v(0) - \int_0^1 u(t)\dot{v}(t)dt.$$

2.6   On the Spaces $H^1(0, 1)$ and $H_0^1(0, 1)$

2.6.1   Definition and Basic Properties

With preparations on absolutely continuous functions from the previous section, we proceed to defining the space in which we will consider our second-order Dirichlet problems.

Definition 2.7 (Spaces $H^1(0, 1)$ and $H_0^1(0, 1)$)
Space $H^1(0, 1)$ consists of absolutely continuous functions $u : [0, 1] \to \mathbb{R}$ whose derivatives (understood a.e. on $[0, 1]$) $\dot{u} : [0, 1] \to \mathbb{R}$ are integrable with square on $[0, 1]$. We define also space $H_0^1(0, 1)$ as follows

$$H_0^1(0, 1) = \left\{ u \in H^1(0, 1) : u(0) = u(1) = 0 \right\}.$$

The space $H^1(0, 1)$ is equipped with the scalar product

$$(u, v)_{H^1} = (u, v)_{L^2} + (\dot{u}, \dot{v})_{L^2} = \int_0^1 u(t)v(t)\,dt + \int_0^1 \dot{u}(t)\dot{v}(t)\,dt \text{ for } u, v \in H^1(0, 1)$$

and the corresponding norm making it into a Hilbert space

$$\|u\|_{H^1} = \sqrt{\|u\|_{L^2}^2 + \|\dot{u}\|_{L^2}^2} = \sqrt{\int_0^1 |u(t)|^2 \, dt + \int_0^1 |\dot{u}(t)|^2 \, dt} \text{ for } u \in H^1(0, 1).$$

Exercise 2.45

Fix a real number $\lambda > 0$. Show that

$$\|u\|_\lambda = \sqrt{\lambda\|u\|_{L^2}^2 + \|\dot{u}\|_{L^2}^2} \quad \text{for } u \in H_0^1(0, 1)$$

is a norm on $H^1(0, 1)$, which is induced by a corresponding inner product. Then show that this norm is equivalent to the standard norm $\|\cdot\|_{H^1}$ on $H^1(0, 1)$.

Exercise 2.46

Show that formula

$$\|u\|_{H^1} = \sqrt{\int_0^1 |u(t)|^2 \, dt} + \sqrt{\int_0^1 |\dot{u}(t)|^2 \, dt} \text{ for } u \in H^1(0, 1)$$

defines an equivalent norm on $H^1(0, 1)$.

We can define

$$H^2(0, 1) = \left\{ u \in H^1(0, 1) \mid \dot{u} \in H^1(0, 1) \right\}$$

with norm for $u \in H^2(0, 1)$

$$\|u\|_{H^2}^2 = \sqrt{\|u\|_{L^2}^2 + \|\dot{u}\|_{L^2}^2 + \|\ddot{u}\|_{L^2}^2}. \tag{2.9}$$

Exercise 2.47

Show that the norm in $H^2(0, 1)$ as given by (2.9) is induced by the following scalar product

$$(u, v)_{H^2} = \int_0^1 u(t) \, v(t) \, dt + \int_0^1 \dot{u}(t) \, \dot{v}(t) \, dt + \int_0^1 \ddot{u}(t) \, \ddot{v}(t) \, dt \text{ for } u, v \in H^2(0, 1).$$

Exercise 2.48

Show that on $H^2(0, 1)$

$$\|u\| = \|u\|_{L^2} + \|\dot{u}\|_{L^2} + \|\ddot{u}\|_{L^2} \text{ for } u \in H^2(0, 1)$$

is an equivalent norm.

▶ **Remark 2.6** Let $u \in H_0^1(0, 1)$. Then we have by a direct calculation and the Hölder Inequality

$$|u(t)| = |u(t) - u(0)| = \left| \int_0^t \dot{u}(s)ds \right| \leq \|\dot{u}\|_{L^1} \leq \|\dot{u}\|_{L^2} \text{ for any } t \in [0, 1].$$

Note that the above inequality is also valid for $u \in \{v \in H^1(0, 1) : v(0) = 0\}$. We also have the obvious relation

$$\|u\|_{L^2} \leq \|u\|_\infty.$$

Theorem 2.5

For any $u \in H_0^1(0, 1)$ there hold the Sobolev *and the* Poincaré *inequalities*

$$\|u\|_\infty \leq \|u\|_{H_0^1} \text{ and } \|u\|_{L^2} \leq \frac{1}{\pi}\|u\|_{H_0^1}. \qquad (2.10)$$

▶ **Remark 2.7** From *the Poincaré Inequality* (2.10) it follows that the norms $\|\cdot\|_{H^1}$ and $\|\cdot\|_{H_0^1}$ are equivalent on $H_0^1(0, 1)$.

The Poincaré constant $\frac{1}{\pi}$ follows directly from the Parseval Identity known form the theory of Fourier expansions, while in Remark 2.6 it follows with some less refined approach but we obtain less accurate constant as well. We will return to finding the Poincaré constant later on, in Sect. 4.6.

Exercise 2.49

Prove that $H_0^1(0, 1)$ is a closed subspace of $H^1(0, 1)$.

Apart from the notion of a norm (strong) convergence, we have also the weak convergence in $H_0^1(0, 1)$, which means as follows: $(u_n) \subset H_0^1(0, 1)$ is weakly convergent to $u_0 \in H_0^1(0, 1)$ if

$$\int_0^1 \dot{u}_n(t)\dot{h}(t)\,dt \to \int_0^1 \dot{u}(t)\dot{h}(t)\,dt \text{ for any } h \in H_0^1(0, 1).$$

In order to investigate the properties of weakly convergent sequences from $H_0^1(0, 1)$, we need to recall the following compactness result, see Theorem 7.25 from [45]. We will provide a more general version in Sect. 5.2 with $[0, 1]$ replaced by a compact metric space.

Theorem 2.6 (Arzela–Ascoli Theorem)

Assume that the sequence of functions $u_n : [0, 1] \to \mathbb{R}$ is uniformly bounded, i.e.,

$$\exists_{M>0} \forall_{t \in [0,1]} \forall_{n \in \mathbb{N}} \, |u_n(t)| \leq M$$

and uniformly equi-continuous, i.e.

$$\forall_{\varepsilon>0} \exists_{\delta>0} \forall_{n \in \mathbb{N}} \forall_{t,s \in [0,1]} \, |t - s| < \delta \implies |u_n(t) - u_n(s)| < \varepsilon.$$

Then sequence (u_n) contains a subsequence which is uniformly convergent on $[0, 1]$.

Theorem 2.7

Let $(u_n) \subset H_0^1(0, 1)$ be weakly convergent to some $u_0 \in H_0^1(0, 1)$. Then $u_n \rightrightarrows u_0$ on $[0, 1]$.

Proof Since $(u_n) \subset H_0^1(0, 1)$ is weakly convergent, it is bounded, i.e., there is some constant $d > 0$ that $\|u_n\|_{H_0^1} \leq d$. Fix $\varepsilon > 0$ and take $\delta < \left(\frac{\varepsilon}{d}\right)^2$. Then we have

$$|u_n(t) - u_n(s)| \leq \int_s^t |\dot{u}_n(\tau)| \, d\tau \leq \sqrt{\int_s^t 1^2 d\tau} \sqrt{\int_s^t |\dot{u}_n(\tau)|^2 \, d\tau}$$
$$= \sqrt{|t - s|} \sqrt{\int_0^1 |\dot{u}_n(\tau)|^2 \, d\tau} < \varepsilon \text{ for } n \in \mathbb{N}, t, s \in [0, 1], \; s < t, \; |t - s| < \delta.$$

By *the Arzela–Ascoli Theorem* sequence (u_n) has a subsequence (u_{n_k}) such that $u_{n_k} \rightrightarrows u_0$. Since any other uniformly convergent subsequence would also approach u_0, we see that the entire sequence (u_n) is uniformly convergent.

Exercise 2.50

Let $(u_n) \subset H_0^1(0, 1)$ be weakly convergent to some $u_0 \in H_0^1(0, 1)$. Show that $u_n \to u_0$ in $L^2(0, 1)$.

Exercise 2.51

Let $(u_n) \subset H_0^1(0, 1)$ be bounded. Show that it is uniformly convergent, up to a subsequence, to some $u_0 \in H_0^1(0, 1)$.

Example 2.6

Let us consider integral functional $F : L^2(0, 1) \to \mathbb{R}$ given by

$$F(u) = \int_0^1 |u(t)|^2 \, dt.$$

Then it is not sequentially weakly continuous according to Example 2.3. Considering the same functional on the space $H_0^1(0, 1)$, we see that now it is sequentially weakly continuous. Indeed, if (u_n) is weakly convergent in $H_0^1(0, 1)$, then it follows from Theorem 2.7 that $u_n \to u_0$ in $L^2(0, 1)$. Now we obtain the assertion since F is obviously continuous on $L^2(0, 1)$.

Example 2.7

Let $G : \mathbb{R} \to \mathbb{R}$ be a continuous function and consider a functional

$$F : H_0^1(0, 1) \to \mathbb{R}$$

given by

$$F(u) = \int_0^1 G(u(t)) \, dt.$$

Then F is sequentially weakly continuous.
Indeed, a sequence (u_n) which is weakly convergent in $H_0^1(0, 1)$ to some u_0 converges also uniformly on $[0, 1]$. The result now follows immediately by the continuity. We can extend easily the assertion of this observation to the case of a jointly continuous function $G : [0, 1] \times \mathbb{R} \to \mathbb{R}$. Note that when considered on $L^2(0, 1)$ functional F would be sequentially weakly lower semicontinuous under additional assumptions that G is convex and that condition (2.11) given below holds for $\alpha \in (1, 2]$.

Let $\mathbb{R}_+ := [0, +\infty)$. Example 2.6 admits a more general version:

Example 2.8

Let $\alpha \geq 1$ be fixed. Assume that $f : [0, 1] \times \mathbb{R} \to \mathbb{R}$ is a Carathéodory function and there exist $\varphi \in L^\infty([0, 1], \mathbb{R}_+)$, $h \in L^1(0, 1)$ such that

$$|f(t, x)| \leq \varphi(t) |x|^\alpha + h(t) \text{ for } a.e.t \in [0, 1] \text{ and all } x \in \mathbb{R}. \tag{2.11}$$

Then functional $F : H_0^1(0, 1) \to \mathbb{R}$ defined by

$$F(u) = \int_0^1 f(t, u(t)) \, dt$$

is sequentially weakly continuous.

Since a sequence (u_n), which is weakly convergent in $H_0^1(0, 1)$ to some u_0 converges also uniformly on $[0, 1]$, there is some $d > 0$ that $\|u_n\|_\infty \leq d$. This further implies that

$$|f(t, u_n(t))| \leq d^\alpha |\varphi(t)| + |h(t)| \quad \text{for } a.e.t \in [0, 1].$$

Hence, applying *the Lebesgue Dominated Convergence Theorem*, we obtain the assertion. Since F is sequentially weakly continuous, it is also continuous.

Exercise 2.52

Find the values of α for which the functional considered in Example 2.8 is well defined and continuous when considered over $L^2(0, 1)$. Find condition under which it is is sequentially weakly continuous there.

We can generalize Example 2.8 and formulate the following lemma whose proof follows alike and is left as an exercise:

Lemma 2.5

Assume that $f : [0, 1] \times \mathbb{R} \to \mathbb{R}$ is an L^1-Carathéodory function. Then functional $F : H_0^1(0, 1) \to \mathbb{R}$ defined by

$$F(u) = \int_0^1 f(t, u(t)) \, dt$$

is sequentially weakly continuous.

Lemma 2.5 will be of importance to us in what follows. We would like to underline that F is not necessarily sequentially weakly continuous when considered on $L^2(0, 1)$.

2.6.2   Embeddings of the Space $H_0^1(0, 1)$

We summarize properties of the space $H_0^1(0, 1)$ concerning the convergence of the sequences into some abstract formalism.

Definition 2.8 (Continuous Imbedding)

Let X and Y be Banach spaces. We say X *is continuously imbedded in* Y and write

$$X \hookrightarrow Y$$

if $X \subset Y$ and if there is a constant $C > 0$ such that

$$\|u\|_Y \leq C\|u\|_X \quad \text{for all } u \in X.$$

Exercise 2.53

Prove that if $X \hookrightarrow Y$ then the identity operator $I : X \to Y$ is bounded.

Definition 2.9 (Compact Imbedding)

Let X, Y be Banach spaces such that X *is continuously imbedded in* Y. We say that X is compactly imbedded in Y and write

$$X \overset{c}{\hookrightarrow} Y$$

if the unit ball in X is precompact in Y (i.e., for every $\varepsilon > 0$ it can be covered by finitely many balls with radius ε) or, equivalently, every bounded sequence in X has a subsequence that converges in Y.

Hence, we can write that $H_0^1(0, 1) \overset{c}{\hookrightarrow} L^2(0, 1)$ and also $H_0^1(0, 1) \overset{c}{\hookrightarrow} C[0, 1]$. We underline that the former will be also of use in the multidimensional case, while the latter in general not.

2.6.3   The Space $H^{-1}(0, 1)$

The dual space of the space $H_0^1(0, 1)$ is denoted by $H^{-1}(0, 1)$, H^{-1} for short. Using *the Riesz Representation Theorem* we identify L^2 with its dual directly. The space $H^{-1}(0, 1)$ which is understood as follows. Let $F \in H^{-1}(0, 1)$, then there exists (not uniquely determined) $f \in L^2(0, 1)$ such that

$$\langle F, u \rangle_{H^{-1}, H_0^1} = \int_0^1 f(t)\, \dot{u}(t)\, dt$$

for all $u \in H_0^1(0, 1)$ and also

$$\|F\|_{H^{-1}} = \|f\|_{L^2}.$$

We have the following continuous and dense inclusions

$$H_0^1(0, 1) \subset L^2(0, 1) \subset H^{-1}(0, 1).$$

The aforementioned says also that any element from $L^2(0, 1)$ can be treated as continuous a linear functional over $H_0^1(0, 1)$.

2.7   On the du Bois-Reymond Lemma and the Regularity Results

We proceed with a lemma which serves as a type of regularity result and which is called du Bois-Reymond Lemma or else the fundamental lemma of the calculus of variation. We give a number of results in the space $H_0^1(0, 1)$. These can be easily shifted to the case of space $C^1[0, 1]$, which we leave as exercises.

> **Lemma 2.6 (Auxiliary Lemma)**
> Let $h \in L^2(0, 1)$ be fixed and let
>
> $$\int_0^1 h(t)\, \dot{v}(t)\, dt = 0 \text{ for all } v \in H_0^1(0, 1).$$
>
> Then there exists a constant $c \in \mathbb{R}$ such that $h(t) = c$ a.e. on $[0, 1]$.

Proof Define function $v_1 : [0, 1] \to \mathbb{R}$ by the following formula

$$v_1(t) = \int_0^t (h(\tau) - c)\, d\tau,$$

where c is chosen so that

$$\int_0^1 (h(\tau) - c)\, d\tau = 0$$

(we see that $c = \int_0^1 h(\tau)\, d\tau$ is an integral mean value of h over $[0, 1]$). Therefore, by Lemma 2.3 we see that function v_1 is absolutely continuous with a derivative equal a.e. on $[0, 1]$ to $h(\cdot) - c$ and $v_1(0) = v_1(1) = 0$. Therefore $v_1 \in H_0^1(0, 1)$. Additionally we see that

$$\int_0^1 c \cdot \dot{v}(t)\, dt = c\,(v(1) - v(0)) = 0 \text{ for any } v \in H_0^1(0, 1).$$

This implies that

$$\int_0^1 h(t)\dot{v}_1(t)\,dt = \int_0^1 (h(t) - c)\dot{v}_1(t)\,dt - \int_0^1 (h(t) - c)^2\,dt$$

Since $(h(t) - c)^2 \geq 0$ for a.e. $t \in [0, 1]$, we see that $h(t) = c$ for a.e. $t \in [0, 1]$.

Lemma 2.7 (du Bois–Reymond Lemma)
Let functions $h \in L^2(0, 1)$, $f \in L^1(0, 1)$ be fixed and assume that

$$\int_0^1 (h(t)\dot{v}(t) + f(t)v(t))\,dt = 0 \text{ for all } v \in H_0^1(0, 1). \qquad (2.12)$$

Then function h is absolutely continuous and $\dot{h}(t) = f(t)$ a.e. on $[0, 1]$.

Proof Note that by Lemma 2.4 function $F : [0, 1] \to \mathbb{R}$ defined by

$$F(t) = \int_0^t f(s)\,ds$$

is absolutely continuous and that $\dot{F}(t) = f(t)$ a.e. on $[0, 1]$. Then we have by Proposition 2.1 that

$$\int_0^1 \dot{F}(t)v(t)\,dt = -\int_0^1 F(t)\dot{v}(t)\,dt$$

and (2.12) has the following equivalent form

$$\int_0^1 (h(t) - F(t))\dot{v}(t)\,dt = 0.$$

By Lemma 2.6 there exists a constant c such that $F(t) = h(t) + c$ for a.e. $t \in [0, 1]$. Now we see that the assertion follows.

Exercise 2.54

Carry out the steps in the proof of Lemma 2.7 in the special case that

$$f(t) = \sin t.$$

What happens when $h(t) = \sin t$?

Exercise 2.55

Formulate conditions under which function h from Lemma 2.7 belongs to $H^2(0, 1)$.

It is obvious now how to formulate the version of the above lemma when $f, h \in C[0, 1]$. We have what follows (recall formula (2.8)):

Corollary 2.2

Let functions $h, f \in C[0, 1]$ *be fixed and assume that*

$$\int_0^1 (h(t)\,\dot{v}(t) + f(t)\,v(t))\,dt = 0 \text{ for all } v \in C_0^1[0, 1].$$

Then $h \in C^1[0, 1]$ *and* $\dot{h}(t) = f(t)$ *on* $[0, 1]$.

We now turn to the so-called *Lagrange Lemma*, which works for continuous functions:

Lemma 2.8 (Lagrange Lemma)

Let $f \in C[0, 1]$ *and assume that*

$$\int_0^1 f(t)\,v(t)\,dt = 0$$

for all $v \in C_0^1[0, 1]$. *Then* $f(t) = 0$ *over* $[0, 1]$.

Proof Suppose there is some $c \in (0, 1)$ that $f(c) > 0$. Then there is an interval $[\alpha, \beta] \subset (0, 1)$ over which f is positive. Consider a test function

$$v(t) = \begin{cases} [(t - \alpha)(\beta - t)]^2, & t \in [\alpha, \beta], \\ 0, & t \notin [\alpha, \beta]. \end{cases} \tag{2.13}$$

Since $v(0) = v(1) = 0$, it follows that $v \in C_0^1[0, 1]$. Moreover $v(t) > 0$ for $t \in (\alpha, \beta)$ and $v \equiv 0$ otherwise. This implies that

$$\int_0^1 f(t)\,v(t)\,dt = \int_\alpha^\beta f(t)\,v(t)\,dt > 0$$

which is a contradiction. We argue similarly assuming that there is some $c \in (0, 1)$ such that $f(c) < 0$.

Exercise 2.56

Show that if $f \in L^1(0, 1)$ is fixed and if

$$\int_0^1 f(t) v(t) \, dt = 0 \text{ for all } v \in H_0^1(0, 1),$$

then $f(t) = 0$ for a.e. $t \in [0, 1]$.

Exercise 2.57

Suppose that $h \in C^1[0, 1]$ and

$$\int_0^1 h(t) v'(t) dt = 0, \quad \text{for all } v \in E_0,$$

where

$$E_0 = v \in \left\{ C^1[0, 1] : v(0) = v'(0) = v(1) = v'(1) = 0 \right\}. \tag{2.14}$$

Use integration by parts and the proof of *the Lagrange Lemma* to conclude that $h = \text{const.}$ on $[0, 1]$. (Do not use a variant of Lemma 2.6).

Exercise 2.58

Assume that $g \in C[0, 1]$ and

$$\int_0^1 \left[g(t) v(t) + h(t) v'(t) \right] dt = 0, \quad \text{for all } v \in E_0,$$

where E_0 is given by (2.14). Argue that $h' = g$ on $[0, 1]$, without using *the du Bois-Reymond Lemma*.

Let $C^0[0, 1] \equiv C[0, 1]$. *The Lagrange Lemma* admits the following higher-order generalization:

Lemma 2.9 (Higher-Order Lagrange Lemma)
Let

$$E_0 = \left\{ v \in C^m[0, 1] : v^{(k)}(0) = v^{(k)}(1) = 0, k = 0, 1, 2, \ldots, m \right\}.$$
(2.15)

If $g \in C[0, 1]$ and for some $m = 0, 1, 2, \ldots$

$$\int_0^1 g(t)v(t)dt = 0, \text{ for all } v \in E_0,$$

then $g \equiv 0$ on $[0, 1]$.

Exercise 2.59

Prove the higher-order Lagrange Lemma. Hint: Suppose $g(c) > 0$ for some $c \in (0, 1)$; take an interval $[\alpha, \beta] \subseteq (0, 1)$ such that $|g(t) - g(c)| \leq g(c)/2$ and define

$$v(t) = \begin{cases} [(t - \alpha)(\beta - t)]^{m+1}, & t \in [\alpha, \beta], \\ 0, & t \notin [\alpha, \beta]. \end{cases}$$

The du Bois–Reymond Lemma also admits higher-order generalization, which is in turn important in consideration of the regularity of solutions to higher-order Dirichlet problems. Although we are not concerned with such problems, we provide the relevant lemma omitting the proof, which is to be found in [49]:

Lemma 2.10
If $h \in C[0, 1]$ and for some $m = 1, 2, \ldots$

$$\int_0^1 h(t)v^{(m+1)}(t)dt = 0, \quad \text{for all } v \in E_0,$$

where E_0 is given by (2.15), then h is a polynomial of degree at most m on $[0, 1]$.

Exercise 2.60

Assume that $h \in C[0, 1]$ and

$$\int_0^1 h(t)v(t)dt = 0, \quad \text{for all } v \in E_0 = \left\{ u \in C[0, 1] : \int_0^1 u(t)dt = 0 \right\}.$$

Show that $h(t) = \text{const.}$ on $[0, 1]$.

Exercise 2.61

Let $\lambda > 0$ be a fixed real number. Show that for $f \in L^2(0, 1)$ and $u \in H_0^1(0, 1)$ the following statements are equivalent:

(a) $u \in H^2(0, 1)$, $\lambda u(t) - \ddot{u}(t) = f(t)$ and $\dot{u}(0) = 0 = \dot{u}(1)$.

(b) $\lambda \int_0^1 u(t)\, v(t)\, dt + \int_0^1 \dot{u}(t)\, \dot{v}(t)\, dt = \int_0^1 f(t)\, v(t)\, dt$ for all $v \in C^1[0, 1]$.

(c) $\lambda \int_0^1 u(t)\, v(t)\, dt + \int_0^1 \dot{u}(t)\, \dot{v}(t)\, dt = \int_0^1 f(t)\, v(t)\, dt$ for all $v \in H_0^1(0, 1)$.

Exercise 2.62

Show that if $u \in H_0^1(0, 1)$ and $v \in C^1[0, 1]$, then $uv \in H_0^1(0, 1)$ with

$$\frac{d}{dt}(uv) = \dot{u}v + u\dot{v}$$

a.e. on $[0, 1]$.

Differentiation in Infinite-Dimensional Spaces 3

In this chapter, we are concerned with differentiation of functionals and next maps in infinite-dimensional spaces. Several notions will be introduced, illustrated, related, and commented. We concentrate mainly on differentiation of functionals, which corresponds to our needs. In this chapter, we follow [4] for theoretical description and [23, 49] for problems and examples. Many theoretical examples are taken following research papers in the area of the calculus of variations.

3.1 The Gâteaux Variation and Its Computation

We start with the easiest notion of the Gâteaux variation and while calculating concrete variations, we will introduce some observations and techniques, which we will use further on. Moreover, we will use the results from this section when we proceed to investigation of another differentiability notions. For technical reasons we introduce the theorem, which we shall use very often in what follows.

Theorem 3.1 (Leibniz Rule)
Assume that functions

$$f, \frac{\partial f}{\partial y} : [a, b] \times [c, d] \to \mathbb{R}$$

are continuous on $[a, b] \times [c, d]$. Then $g : [c, d] \to \mathbb{R}$ defined by formula

$$g(y) = \int_a^b f(x, y)\, dx$$

(Continued)

© The Author(s), under exclusive license to Springer Nature Switzerland AG 2024
M. Galewski, *Basics of Nonlinear Optimization*, Compact Textbooks in
Mathematics, https://doi.org/10.1007/978-3-031-77160-6_3

Theorem 3.1 (continued)

is continuously differentiable on $[c, d]$ *and the following formula holds*

$$g'(y) = \int_a^b \frac{\partial f(x, y)}{\partial y} dx \text{ for } y \in [c, d].$$

Example 3.1

Let $d > 1$. We see that functions $g_1, g_2 : [1, d] \to \mathbb{R}$ given by

$$g_1(y) = \int_0^1 \arctan \frac{x}{y} dx \text{ and } g_2(y) = \int_0^1 \ln\left(x^2 + y^2\right) dx$$

are continuously differentiable and for $y \in [1, d]$ we have

$$g_1'(y) = \int_0^1 \frac{\partial}{\partial y}\left(\arctan \frac{x}{y}\right) dx = -\int_0^1 \frac{x}{x^2 + y^2} dx = \frac{1}{2} \ln\left(\frac{y^2}{1 + y^2}\right),$$

$$g_1'(y) = \int_0^1 \frac{\partial}{\partial y} \ln\left(x^2 + y^2\right) dx = \int_0^1 \frac{2y}{x^2 + y^2} dx = 2 \arctan \frac{1}{y}.$$

Definition 3.1 (First Gâteaux Variation)

Let E be a real linear space. Let $x_0 \in E$ be fixed. The functional $F : E \to \mathbb{R}$ has *first Gâteaux variation* at $x_0 \in E$ in direction $h \in E$ provided the following limit exists

$$\lim_{\varepsilon \to 0} \frac{F(x_0 + \varepsilon h) - F(x_0)}{\varepsilon} \tag{3.1}$$

which we denote by $\delta F(x_0, h)$.

We will write Gâteaux variation for short meaning of the first Gâteaux variation. Note that the limit in (3.1) is taken in $\mathbb{R}$ which means that the Gâteaux variation can be defined in case E is a linear space. Note also that the existence of the limit in the aforementioned definition is equivalent to the existence of the derivative of a real-valued function $\varepsilon \mapsto F(x_0 + \varepsilon h)$ at $\varepsilon_0 = 0$ and thus it can be calculated provided that:

(a) $F(x_0)$ is defined;
(b) $F(x_0 + \varepsilon h)$ is defined for all sufficiently small $\varepsilon > 0$; and then

$$\delta F(x_0; h) = \frac{\partial}{\partial \varepsilon} F(x_0 + \varepsilon h)\Big|_{\varepsilon=0}.$$

The Gâteaux variation is obviously uniquely defined, although it works only for functionals, which means that we do not have the counterpart of this notion for maps between two infinite-dimensional spaces. The Gâteaux variation of F at x_0 depends only on the local behavior of F on a line passing through x_0 in direction h. Such a variation need not exist in any direction h, or it may exist in some directions and not in others as is the case with the directional derivative known from calculus courses.

▶ **Remark 3.1** As a direct consequence of (3.1) and the linearity of the ordinary derivative, we see that the Gâteaux variation is a linear operation. Namely, if at some $x_0 \in E$, functionals F and $\tilde{F}$ each have the Gâteaux variations in the same direction h, then for constants $c, \tilde{c} \in \mathbb{R}$, we see that $\delta(cF + \tilde{c}\tilde{F})(x_0; h)$ exists and equals $c\delta F(x_0, h) + \tilde{c}\delta\tilde{F}(x_0, h)$. We also see that $\delta F(x_0; 0_E) = 0$.

The following are direct properties of the Gâteaux variation that follow from the definition. We provide them as an exercise for the reader to work on (by consulting any calculus text, for example [11, 12]) before moving further:

Exercise 3.1

Let E be a real linear space. Assume that functionals $F, G : E \rightarrow \mathbb{R}$ have first Gâteaux variations at $x_0 \in E$ in the same direction $h \in E$.

(a) Verify the product formula

$$\delta(FG)(x_0; h) = \delta F(x_0; h)G(x_0) + F(x_0)\delta G(x_0; h).$$

(b) Establish the quotient rule

$$\delta\left(\frac{F}{G}\right)(x_0; h) = \frac{\delta F(x_0; h)G(x_0) - F(x_0)\delta G(x_0; h)}{(G(x_0))^2},$$

provided that $G(x_0) \neq 0$.

(c) Supposing that $g \in C^1(\mathbb{R})$, show that (*the Chain Rule*)

$$\delta(g(F))(x_0; h) = g'(F(x_0))\delta F(x_0; h).$$

Exercise 3.2

Let E be a real linear space and consider functionals $F, \tilde{F} : E \to \mathbb{R}$. Assume that for some $x_0, h, \tilde{h} \in E$, there exist $\delta F(x_0; h)$, $\delta F(x_0; \tilde{h})$ and $\delta \tilde{F}(x_0; h)$ and that $c \in \mathbb{R}$. Demonstrate that:

(a) $\delta(cF)(x_0; h) = \delta F(x_0; ch) = c\delta F(x_0; h)$.
(b) $\delta(F + \tilde{F})(x_0; h) = \delta F(x_0; h) + \delta \tilde{F}(x_0; h)$.
(c) Check whether necessarily

$$\delta F(x_0; h + \tilde{h}) = \delta F(x_0; h) + \delta F(x_0; \tilde{h})?$$

We warn the reader about the chain rules that appear in this chapter. These comply with the following general scheme: the outer function must rather be C^1, then the composition enjoys at least the same differentiability property as the inner map.

Now we proceed to examples concerning computation of the Gâteaux variation:

Example 3.2

If $F \in C^1(\mathbb{R}^N)$ and $x_0, h \in \mathbb{R}^N$, then $\delta F(x_0; h)$ is just the directional derivative and thus

$$\delta F(x_0; h) = \lim_{\varepsilon \to 0} \frac{F(x_0 + \varepsilon h) - F(x_0)}{\varepsilon} = \nabla F(x_0) \cdot h.$$

The aforementioned example suggests alternative way of calculating the Gâteaux variation:

▶ **Remark 3.2** Let E be real linear space and $x_0, h \in E$, $\varepsilon \neq 0$ be fixed. Let $F : E \to \mathbb{R}$. Assume that the numerical function

$$\varepsilon \to \frac{\partial}{\partial \varepsilon} F(x_0 + \varepsilon h)$$

is continuous at least at $\varepsilon_0 = 0$. Then instead of using formula (3.1) we differentiate the numerical function $\varepsilon \to F(x_0 + \varepsilon h)$ and evaluate the result at $\varepsilon_0 = 0$.

In what follows, we will differentiate functionals defined on $C[0, 1]$, $C^1[0, 1]$ and also on $H_0^1(0, 1)$.

Example 3.3

Put $E = C[0, 1]$ and define $F : E \to \mathbb{R}$ as follows

$$F(u) = \int_0^1 \left[\sin^3 t + u^2(t) \right] dt \text{ for } u \in E.$$

Due to the continuity, we understand the integral in the sense of Riemann and we see that F is well defined, i.e., finite everywhere. Thus, for fixed $u_0, h \in E$ and $\varepsilon \neq 0$ we see that $u_0 + \varepsilon h \in E$ (as E is a linear space). Hence after successive cancellations we obtain the following ratio

$$\frac{F(u_0 + \varepsilon h) - F(u_0)}{\varepsilon} = \frac{1}{\varepsilon} \int_0^1 \left[(u_0 + \varepsilon h)^2(t) - u_0^2(t) \right] dt =$$
$$\frac{1}{\varepsilon} \int_0^1 \left[u_0^2(t) + 2\varepsilon u_0(t) h(t) + \varepsilon^2 h^2(t) - u_0^2(t) \right] dt =$$
$$2 \int_0^1 u_0(t) h(t) dt + \varepsilon \int_0^1 h^2(t) dt.$$

The aforementioned ratio is a numerical function of ε and both integrals in the last expression are finite. Taking the limit as $\varepsilon \to 0$ we see that for any $u_0, h \in E$

$$\delta F(u_0; h) = 2 \int_0^1 u_0(t) h(t) dt. \tag{3.2}$$

Now we calculate (3.2) using the method described in Remark 3.2. We see that

$$F(u_0 + \varepsilon h) = \int_0^1 \left[\sin^3 t + u_0^2(t) \right] dt + 2\varepsilon \int_0^1 u_0(t) h(t) dt + \varepsilon^2 \int_0^1 h^2(t) dt.$$

Hence, we arrive at once at formula (3.2).

Example 3.4

Let $\rho \in C[0, 1]$ be fixed. Put $E = C^1[0, 1]$ and define $F : E \to \mathbb{R}$ as follows

$$F(u) = \int_0^1 \rho(t) \sqrt{1 + (u'(t))^2} dt \text{ for } u \in E.$$

Using the method described in Remark 3.2 and *the Leibniz Rule* we obtain for fixed $u, h \in E$ and $\varepsilon \in \mathbb{R}$ that

$$\frac{\partial}{\partial \varepsilon} F(u + \varepsilon h) = \int_0^1 \frac{\partial}{\partial \varepsilon} \left[\rho(t) \sqrt{1 + ((u + \varepsilon h)'(t))^2} \right] dt =$$
$$\int_0^1 \frac{\rho(t)(u + \varepsilon h)'(t) h'(t)}{\sqrt{1 + ((u + \varepsilon h)'(t))^2}} dt$$

and evaluating at $\varepsilon = 0$, we finally get

$$\delta F(u; h) = \int_0^1 \frac{\rho(t)u'(t)}{\sqrt{1 + u'(t)^2}} h'(t) dt.$$

We have a number of standard exercises, which will help us in a better understanding of the variation:

Exercise 3.3

Put $E = C^1[0, 1]$, $F : E \to \mathbb{R}$. Compute the Gâteaux variation when:

(a) $F(u) = (u(0))^3$.

(b) $F(u) = \int_a^b \left[(u(t))^3 + t\left(u'(t)\right)^2 \right] dt$.

(c) $F(u) = \int_0^1 \sqrt{2 + t^2} - \sin u'(t) dt$.

(d) $F(u) = \int_0^1 \left[e^t u(t) - 3\left(u'(t)\right)^4 \right] dt + 2\left(u'(0)\right)^2$.

(e) $F(u) = \int_0^1 \left[t^2 (u(t))^2 + e^{u'(t)} \right] dt$.

(f) $F(u) = \sin u'(0) + \cos u(1)$.

(g) $F(u) = \left(\int_0^1 \left[2u'(t) + t^2 u(t) \right] dt \right) \left(\int_0^1 \left[1 + u'(t) \right]^2 dt \right)$.

(h) $F(u) = \int_0^1 u(t)dt / \int_0^1 \left[1 + u'(t)^2 \right] dt$.

▶ **Remark 3.3** Situation changes when we aim at considering the action functional F over $L^2(0, 1)$ or over $H_0^1(0, 1)$. Since the limit is taken over the real line, the Gâteaux variation in Examples 3.3 and 3.4 will have similar form if we replace $C[0, 1]$ with $L^2(0, 1)$ in Example 3.3 and $C^1[0, 1]$ with $H_0^1(0, 1)$ in Example 3.4. However, it will require different tools to compute the Gâteaux variation in $H_0^1(0, 1)$. This why we will directly use the Lebesgue Dominated Convergence Theorem in examples that follow.

Exercise 3.4

Put $E = L^2(0, 1)$ and define $F : E \to \mathbb{R}$ as follows

$$F(u) = \int_0^1 \left[\cos^3 t + u(t) \right] dt \text{ for } u \in E.$$

Show that F is well defined and compute its Gâteaux variation.

Example 3.5

Let $\rho \in L^2(0, 1)$ be fixed. We put $E = H_0^1(0, 1)$ and define $F : E \to \mathbb{R}$ as in the above by

$$F(u) = \int_0^1 \rho(t)\sqrt{1 + (\dot{u}(t))^2}\,dt \text{ for } u \in H_0^1(0, 1).$$

We will prove that F is well defined, and next we will calculate its Gâteaux variation (at each point in every direction). Indeed, for any fixed $u \in H_0^1(0, 1)$ we have what follows, where we use some direct decomposition and also the Hölder inequality:

$$\int_0^1 \rho(t)\sqrt{1 + (\dot{u}(t))^2}\,dt \le$$
$$\int_{\{t:|\dot{u}(t)|\le 1\}} |\rho(t)|\sqrt{1 + (\dot{u}(t))^2}\,dt + \int_{\{t:|\dot{u}(t)|>1\}} |\rho(t)|\sqrt{1 + (\dot{u}(t))^2}\,dt \le$$
$$\sqrt{2}\int_{\{t:|\dot{u}(t)|\le 1\}} |\rho(t)|\,dt + \sqrt{2}\int_{\{t:|\dot{u}(t)|>1\}} |\rho(t)|\,|\dot{u}(t)|\,dt \le$$
$$\sqrt{2}\left(\|\rho\|_{L^2} + \|\rho\|_{L^2}\,\|\dot{u}\|_{L^2}\right).$$

In order to compute the Gâteaux variation, we will employ directly the definition. Applying *the Lagrange Mean Value Theorem* to a C^1 function $x \mapsto \sqrt{1 + x^2}$ we have the following estimation for any $a, b \in \mathbb{R}$

$$\left|\sqrt{1 + a^2} - \sqrt{1 + b^2}\right| \le \frac{c}{\sqrt{1 + c^2}}\,|a - b| \le |a - b| \text{ where } c \in (a, b).$$

Therefore, for fixed $u, h \in E$ and $\varepsilon \in \mathbb{R}$ we obtain for a.e. $t \in [0, 1]$:

$$\frac{1}{\varepsilon}\rho(t)\left(\sqrt{1 + \left(\frac{d}{dt}(u + \varepsilon h)(t)\right)^2} - \sqrt{1 + (\dot{u}(t))^2}\right) \le \rho(t)\dot{h}(t)$$

and we see that function $t \to \rho(t)\dot{h}(t)$ is integrable over $[0, 1]$. Moreover :

$$\lim_{\varepsilon \to 0} \frac{\rho(t)\left(\sqrt{1 + \left(\frac{d}{dt}(u + \varepsilon h)(t)\right)^2} - \sqrt{1 + (\dot{u}(t))^2}\right)}{\varepsilon}$$
$$- \frac{\rho(t)\dot{u}(t)\dot{h}(t)}{\sqrt{1 + (\dot{u}(t))^2}} \text{ for a.e. } t \subset [0, 1].$$

In a consequence by *the Lebesgue Dominated Convergence Theorem* we get

$$\lim_{\varepsilon \to 0} \frac{F(u + \varepsilon h) - F(u)}{\varepsilon} = \int_0^1 \frac{\rho(t)\dot{u}(t)\dot{h}(t)}{\sqrt{1 + (\dot{u}(t))^2}}\,dt \text{ for a.e. } t \in [0, 1].$$

Exercise 3.5

Put $E = L^2(0, 1)$ and define $F : E \to \mathbb{R}$ as follows:

(a)

$$F(u) = \int_0^1 \cos(u(t))\, dt \text{ for } u \in E.$$

(b) For $\theta \in (1, 2)$

$$F(u) = \int_0^1 \left[\arctan(u(t)) + (u(t))^\theta\right] dt \text{ for } u \in E.$$

Show that F is well defined and compute its Gâteaux variation at any point $u \in E$ in every direction $h \in E$.

Example 3.6

We investigate the following functional

$$F(u) = \int_0^1 \sin u(t) dt + u^2(1).$$

Since F contains the evaluation term $u^2(1)$ it does not make sense to consider it in $L^2(0, 1)$. Since any function from $H_0^1(0, 1)$ is absolutely continuous on $[0, 1]$, we may consider F in $E = H_0^1(0, 1)$, where it is well defined. We see for any fixed $u, h \in E$ and $\varepsilon \in \mathbb{R}$:

$$F(u + \varepsilon h) = \int_0^1 \sin(u(t) + \varepsilon h(t)) dt + (u + \varepsilon h)^2(1).$$

Since the integrand is a C^1 function of the parameter, we can apply *the Leibniz Rule* in order to conclude that

$$\delta F(u; h) = \int_0^1 [\cos u(t) h(t) dt + 2u(1)h(1).$$

Now we proceed to some general example, which we also consider on both settings, i.e., in the space $C^1[0, 1]$ and in the Sobolev space $H_0^1(0, 1)$.

Example 3.7

Let $f \in C\left([0, 1] \times \mathbb{R}^2\right)$. Then the functional

$$F(u) = \int_0^1 f\left(t, u(t), u'(t)\right) dt = \int_0^1 f[u(t)] dt, \tag{3.3}$$

is well defined over $E = C^1[0, 1]$ without any additional assumptions on the integrand. In order to apply *the Leibniz Rule* for the auxiliary function (with fixed $u, h \in E$)

$$F(u + \varepsilon h) = \int_0^1 f\left(t, (u + \varepsilon h)(t), (u + \varepsilon h)'(t)\right) dt = \int_0^1 f[(u + \varepsilon h)(t)]dt,$$

we must know that the partial derivatives f_u, f_z of f with respect to the second and third variable exist and are jointly continuous on $[0, 1] \times \mathbb{R}^2$. Having assumed these conditions we see what follows by integration under the integral sign

$$\frac{\partial}{\partial \varepsilon} F(u + \varepsilon h) = \int_0^1 \frac{\partial}{\partial \varepsilon} f[(u + \varepsilon h)(u)]dt =$$
$$\int_0^1 \left(f_u[(u + \varepsilon h)(t)]h(t) + f_z[(u + \varepsilon h)(t)]h'(t)\right) dt$$

Thus, evaluating the aforementioned at $\varepsilon = 0$ we obtain

$$\delta F(u; h) = \int_0^1 \left(f_u[u(t)]h(t) + f_z[u(t)]h'(t)\right) dt \text{ for all } u, h \in E.$$

▶ **Remark 3.4** The most common type of integrand $f : [0, 1] \times \mathbb{R}^2 \to \mathbb{R}$ which appears in Example 3.7 is the following one

$$f(t, x, z) = g(t, x) + \frac{1}{2} |z|^2,$$

where $g \in C([0, 1] \times \mathbb{R})$ is such that $g_u \in C([0, 1] \times \mathbb{R})$. Then functional $F : E \to \mathbb{R}, E = C^1[0, 1]$, given by formula (3.3), reads

$$F(u) = \int_0^1 g(t, u(t)) dt + \frac{1}{2} \int_0^1 |u'(t)|^2 dt \tag{3.4}$$

with the Gâteaux variation

$$\delta F(u; h) = \int_0^1 (g_u(t, u(t))h(t) + u'(t)h'(t)) dt \text{ for all } u, h \in E.$$

Another example is

$$F(u) = \int_0^1 \left(\rho(t)\sqrt{1 + |u'(t)|^2} + g(t, u(t))\right) dt$$

for some fixed $\rho \in C[0, 1]$. We will be preoccupied with such integrands in the sequel.

▶ **Remark 3.5** As mentioned, we cannot shift the results given in Example 3.7 to the case of a functional defined on $H_0^1(0, 1)$ without any additional growth conditions. We note that the joint continuity of the integrand is not sufficient for F given by (3.3) to be well defined over $H_0^1(0, 1)$. For example when

$$f(t, x, z) = x^5 + z^5$$

we note that the term

$$\int_0^1 u^5(t) dt$$

is finite for each $u \in H_0^1(0, 1)$. Indeed, recall that any function from $H_0^1(0, 1)$ is integrable with any power $p > 1$. We have the following well-known inequality for $p_2 > p_1 > 1$:

$$\int_0^1 |u(t)|^{p_1} dt \leq \left(\int_0^1 |u(t)|^{p_2} dt \right)^{p_1/p_2}$$

valid on $L^{p_2}(0, 1)$.

On the other hand, the term

$$\int_0^1 (\dot{u}(t))^5 dt$$

need not finite be since we know that $\dot{u}$ is integrable with square.

Now we proceed to some version of Example 3.7, which can be considered in $H_0^1(0, 1)$, namely we will investigate functional like (3.4) with suitable assumptions on g.

Example 3.8

Put $E = H_0^1(0, 1)$. Let $g : [0, 1] \times \mathbb{R} \to \mathbb{R}$ be an L^1–Carathéodory function such that the partial derivative g_u of g with respect to the second variable exists and $g_u : [0, 1] \times \mathbb{R} \to \mathbb{R}$ is an L^1–Carathéodory function as well. Then functional $F : E \to \mathbb{R}$ defined by (3.4) has the Gâteaux variation at any point $u \in E$ in every direction $v \in E$ and

$$\delta F(u; v) = \int_0^1 (g_u(t, u(t))v(t) + \dot{u}(t)\dot{v}(t)) dt. \tag{3.5}$$

We see that both functional F and the term $\int_0^1 g_u(t, u(t))v(t)dt$ are well defined, which follows since g and g_u are L^1–Carathéodory. Indeed, since u is also absolutely continuous, we find some $d > 0$ that $|u(t)| \leq d$ for all $t \in [0, 1]$.

Then function $t \mapsto \max_{|x| \le d} |g(t, x)|$ belongs to $L^1(0, 1)$. Similarly $t \mapsto \max_{|z| \le d} |g_u(t, z)|$ belongs to $L^1(0, 1)$.

Now we proceed to justification the formula for the Gâteaux variation (3.5). The application of *the Lagrange Mean Value Theorem* says that for $\varepsilon > 0$

$$\left| \frac{g(t, z + \varepsilon w) - g(t, z)}{\varepsilon} \right| \le \max_{c \in [z, w]} |g_u(t, c)| \, |w| \quad \text{for } a.e.t \in [0, 1] \text{ and any } z, w \in \mathbb{R}$$

The aforementioned estimation together with the assumption that g_u is an L^1–Carathéodory function allows us to apply *the Lebesgue Dominated Convergence Theorem* in order to get the assertion.

Exercise 3.6

Put $E = H_0^1(0, 1)$. Show that if we assume $g, g_u : [0, 1] \times \mathbb{R} \to \mathbb{R}$ to be (jointly) continuous then functional $F : E \to \mathbb{R}$ defined by (3.4) has the Gâteaux variation at any point $x \in E$ in every direction $h \in E$ defined by (3.5).

We conclude this section with some additional definitions.

Definition 3.2 (*n*-th Gâteaux Variation)

Let E be a real linear space and $x_0 \in E$. Let $F : E \to \mathbb{R}$ be a functional and let $h \in E$ be fixed. Set $g : \mathbb{R} \to \mathbb{R}$ by

$$g(t) = F(x_0 + th).$$

Functional F is said to have the n–th Gâteaux variation at x_0 in direction h if the real function g is n–times differentiable at $t_0 = 0$. When F has n–th Gâteaux variation in any direction at x_0, we say that F has n–th Gâteaux variation at x_0. We denote the n–th Gâteaux variation in direction h at point x_0 by

$$F^{(n)}(x_0; h).$$

We will still write $\delta F(x_0; h)$ instead of $F^{(1)}(x_0; h)$.

Example 3.9

Assume that E is a real Hilbert space. We consider the functional $F : E \to \mathbb{R}$ given by

$$F(x) = \frac{1}{2}(x, x)_E = \frac{1}{2} \|x\|^2.$$

Fix $x_0 \in E$ and take any direction $h \in E$. We define $g : \mathbb{R} \to \mathbb{R}$ given by

$$g(t) = F(x_0 + th)$$

and observe that

$$g(t) = (x_0 + th, x_0 + th)_E = \frac{1}{2} \|x_0\|^2 + t(x_0, h)_E + t^2 \frac{1}{2} \|h\|^2$$

which means that g is of class C^2 on $\mathbb{R}$ and therefore

$$\delta F(x_0; h) = g'(0) = (x_0, h)_E \text{ and } F^{(2)}(x_0; h) = g''(0) = (h, h)_E.$$

We see also that $F^{(n)}(x_0; h) = 0$ for $n \geq 3$.

3.2   On the Fermat Rule

We start with some restriction on the Gâteaux variation, which is suitable when looking for minimizers.

Definition 3.3 (Lagrange Variation)

Let E be a real linear space and $x_0 \in E$ be fixed. The functional $F : E \to \mathbb{R}$ has first Lagrange variation at x_0 in direction $h \in E$ provided the following limit exists

$$\lim_{\lambda \to 0^+} \frac{F(x_0 + \lambda h) - F(x_0)}{\lambda}$$

which we denote by $F'(x_0; h)$.

Note that we can consider the differentiability of functionals defined over certain set in E as well. In this case one must make sure that it makes sense to move along a given direction at least as $\lambda \to 0^+$. It is also obvious that any functional that has the Gâteaux variation also has the Lagrange variation and both coincide. Nevertheless there is a large class of functions, namely convex functions, which have the Lagrange variation without any other additional assumptions about differentiability. The exact formulation will be given after some technical lemma:

Lemma 3.1

Let E be a real linear space. Assume that $F : E \to \mathbb{R}$ is a convex functional. Then for each $x_0, h \in E$ the function $\varphi : (0, +\infty) \to \mathbb{R}$ defined by the formula

$$\varphi(\lambda) = \frac{F(x_0 + \lambda h) - F(x_0)}{\lambda}$$

is nondecreasing.

Proof Let $0 < s \le t$. Then for any $x_0, h \in E$ we obtain

$$F(x_0 + sh) - F(x_0) = F(\tfrac{s}{t}(x_0 + th) + \tfrac{t-s}{t}x_0) - F(x_0) \le$$
$$\le \tfrac{s}{t}F(x_0 + th) + \tfrac{t-s}{t}F(x_0) - F(x_0) =$$
$$\tfrac{s}{t}(F(x_0 + th) - F(x_0)).$$

The aforementioned immediately implies the assertion, i.e.

$$\varphi(s) \le \varphi(t).$$

From Lemma 3.1 and from the well-known result concerning the existence of a one-sided limit for monotone real-valued function defined on the real line, we obtain what follows (we leave the proof as an easy exercise):

Lemma 3.2
Let E be a real linear space. Assume that $F : E \to \mathbb{R}$ is a convex functional. Then at each $x_0 \in E$ there exists the Lagrange variation in any direction.

Theorem 3.2 (Fermat Rule)
Let E be a real linear space. Let $\emptyset \neq S \subset E$ and let $F : S \to \mathbb{R}$. Then the following holds:

(a) *Let $x_0 \in S$ be a minimizer of F over S. If functional F has the Lagrange variation at x_0 in direction $x - x_0$ for each $x \in S$ then*

$$F'(x_0, x - x_0) \ge 0 \text{ for all } x \in S. \tag{3.6}$$

(b) *If S is additionally convex and if F is a convex functional which has the Lagrange variation at $x_0 \in S$ in direction $x - x_0$ for each $x \in S$ and if (3.6) is satisfied, then x_0 is a minimizer of F over S.*

(c) *If $S = E$, F has the Lagrange variation at $x_0 \in E$ in each direction $h \in E$ and if x_0 is a minimizer of F over E, then*

$$F'(x_0, h) = 0.$$

Proof

(a) Let $x \in S$. Since x_0 is a minimizer of F over S and since F has the Lagrange variation at x_0 in direction $x - x_0$, it follows

$$\frac{F(x_0 + \lambda(x - x_0)) - F(x_0)}{\lambda} \ge 0.$$

for sufficiently small $\lambda \in \mathbb{R}_+$ such that $x_0 + \lambda(x - x_0) \in S$. Letting $\lambda \to 0^+$ we see from the above

$$F'(x_0, x - x_0) \geq 0.$$

(b) From the assumptions for all $\lambda \in (0, 1]$ and all $x \in S$ we have

$$F(x_0 + \lambda(x - x_0)) \leq (1 - \lambda)F(x_0) + \lambda f(x)$$

or equivalently

$$F(x_0 + \lambda(x - x_0)) - F(x_0) \leq \lambda \left(F(x) - F(x_0) \right).$$

Letting $\lambda \to 0^+$ we now see that

$$0 \leq F'(x_0, (x - x_0)) \leq F(x) - F(x_0).$$

Which means that x_0 is a minimizer of F over S since $x \in S$ is arbitrarily chosen.

(c) Since $E = S$ we fix any $h \in E$ and consider function $g : \mathbb{R} \to \mathbb{R}$ defined by $g(t) = F(x_0 + th)$, which has a minimizer at 0. But this implies that

$$F'(x_0, h) = 0.$$

We need to assume the existence of the variation in condition (b) due to the fact that set S need not be the whole space.

Example 3.10

We would like to mention the following function $F : \mathbb{R} \times \mathbb{R} \to \mathbb{R}$

$$F(x, y) = \begin{cases} x^2 \left(1 + \frac{1}{y}\right), & y \neq 0, \\ 0, & y = 0. \end{cases}$$

which is not continuous at $(0, 0)$ and whose first Lagrange variation reads

$$F'(0, h) = \begin{cases} \frac{(h_1)^2}{h_2}, & h_2 \neq 0, \\ 0, & h_2 = 0. \end{cases}$$

Note that this variation is not linear in h, which restricts the applicability of Theorem 3.2.

Let E be a real linear space. Let $x_0 \in E$ be a local minimizer of a functional $F : E \to \mathbb{R}$. Assume that the n-th Gateaux variation of F is defined at this point. Prove that there exists an integer l with $2l \leqslant n$ such that

$$\delta F (x_0; h) = 0, \ldots, F^{(2l-1)} (x_0; h) = 0 \text{ and } F^{(2l)} (x_0; h) \geqslant 0$$

for every $h \in E$. Hint: The proof follows at once from the definition of the n-th variation and from the corresponding theorem for functions of one variable involving higher-order test.

3.3 The Gâteaux Derivative

Now we turn to the situation when the Gâteaux variation is a linear mapping of the direction. This is in compliance with what we know from differentiation of functions of several variables.

Definition 3.4 (Gâteaux Derivative)

Let E be a real Banach space. A functional $F : E \to \mathbb{R}$ is said to be Gâteaux differentiable at $x_0 \in E$ if there exists a bounded (i.e., sending bounded sets into bounded sets) linear functional $F'(x_0) \in E^*$ such that for every $h \in E$

$$\lim_{t \to 0} \frac{F(x_0 + th) - F(x_0)}{t} = \langle F'(x_0), h \rangle. \tag{3.7}$$

The functional $F'(x_0)$ is then called the Gâteaux derivative of F at x_0.

▶ **Remark 3.6** Condition (3.7) in the definition of Gâteaux derivative can be written (in each fixed direction h) as follows

$$F(x_0 + th) - F(x) - t \langle F'(x_0), h \rangle = o(t),$$

where o is the Landau symbol, i.e. $\lim_{t \to 0} \frac{o(t)}{t} = 0$.

We have the following properties of the Gâteaux derivative:

(a) *When $F : \mathbb{R} \to \mathbb{R}$, then the Gâteaux derivative coincides with the classical derivative;*
(b) *When $F : \mathbb{R}^N \to \mathbb{R}$, then the existence of the Gâteaux derivative is implied by the existence and continuity of all partial derivatives;*
(c) *The existence of Gâteaux derivative implies the existence of the Gâteaux variation and that both are equal (but not the vice-versa);*
(d) *The existence of Gâteaux derivative does not imply neither the continuity nor the lower semicontinuity.*

We start with demonstrating how to compute the Gâteaux derivative following the steps:

(a) find the *Gâteaux variation*,
(b) prove that the *Gâteaux variation* is a bounded (and therefore continuous) linear mapping.

Example 3.11

Let $E = C[0, 1]$ and let us consider the functional $F : E \to \mathbb{R}$ given by

$$F(u) = \int_0^1 \frac{\sin u(s)}{1 + u^2(s)} ds.$$

Let $u \in C[0, 1]$ be fixed and let us also fix a direction $h \in E$. We define the following auxiliary function $g : \mathbb{R} \to \mathbb{R}$ by

$$g(t) = \int_0^1 \frac{\sin(u(s) + th(s))}{1 + (u(s) + th(s))^2} ds.$$

Using *the Leibniz Rule* we arrive at

$$g'(t) = \int_0^1 \frac{\cos(u(s) + th(s)) h(s) \left(1 + (u(s) + th(s))^2\right)}{\left(1 + (u(s) + th(s))^2\right)^2} ds$$
$$- \int_0^1 \frac{2 \sin(u(s) + th(s)) (u(s) + th(s)) h(s)}{\left(1 + (u(s) + th(s))^2\right)^2} ds.$$

Since g' is continuous at $t_0 = 0$, we have

$$g'(0) = \int_0^1 \frac{\cos u(s) \left(1 + (u^2(s))\right) - 2 \sin u(s) u(s)}{\left(1 + u^2(s)\right)^2} h(s) ds.$$

The linearity of the aforementioned formula in h is obvious. Now we prove the continuity. Let us denote for $s \in \mathbb{R}$

$$\varphi(s) = \frac{\cos u(s) \left(1 + (u^2(s))\right) - 2 \sin u(s) u(s)}{\left(1 + u^2(s)\right)^2}.$$

Let $h_n \rightrightarrows h$ on $[0, 1]$. Then by the well known properties of *the Riemann Integral*:

$$\lim_{n \to +\infty} \int_0^1 \varphi(s) h_n(s) ds = \int_0^1 \varphi(s) \left(\lim_{n \to +\infty} h(s)\right) ds = \int_0^1 \varphi(s) h(s) ds.$$

Since the aforementioned formula defines a bounded linear functional (of variable h), the Gâteaux differentiability (at any point) is evident.

Example 3.12

Assume that E is a real Hilbert space. We consider $F : E \to \mathbb{R}$ given by

$$F(x) = \frac{1}{2}(x, x)_E$$

for which we have computed the Gâteaux variation at each $x \in E$ in any direction $h \in E$

$$h \mapsto (x, h)_E$$

which by *the Riesz Representation Theorem* defines the linear continuous functional on E.

Example 3.13

Put $E = H_0^1(0, 1)$. Let $g : [0, 1] \times \mathbb{R} \to \mathbb{R}$ be an L^1−Carathéodory function such that the partial derivative $g_u : [0, 1] \times \mathbb{R} \to \mathbb{R}$ of g with respect to the second variable exists is an L^1−Carathéodory function as well. Consider $F : E \to \mathbb{R}$, given by formula (3.3), i.e.

$$F(u) = \int_0^1 g(t, u(t))\, dt + \frac{1}{2} \int_0^1 |\dot{u}(t)|^2\, dt.$$

We know from Example 3.8 that F has the Gâteaux variation defined by the following formula

$$\delta F(u; h) = \int_0^1 g_u(t, u(t))\, h(t)\, dt + \int_0^1 \dot{u}(t)\dot{h}(t)dt \text{ for } u, h \in E.$$

It is immediate to note that

$$h \to \int_0^1 g_u(t, u(t))\, h(t)\, dt$$

defines a linear bounded functional on $H_0^1(0, 1)$. Indeed, by *the Sobolev* and *the Poincaré Inequality* we have

$$\int_0^1 |g_u(t, u(t))\, h(t)|\, dt \leq \int_0^1 |g_u(t, u(t))|\, dt \cdot \|h\|_\infty \leq g_1 \|h\|_{H_0^1}, \tag{3.8}$$

where $g_1 = \int_0^1 |g_u(t, u(t))|\, dt$ is a finite number since g_u is an L^1−Carathéodory function. Thus we reach the conclusion that F is Gâteaux differentiable over E.

Exercise 3.8

Assume in Example 3.13 that g_u is an L^2–Carathéodory function. Use the Schwarz Inequality in proving 3.8).

Exercise 3.9

Prove that the second functional given in Remark 3.4 is Gâteaux differentiable.

Exercise 3.10

Prove that a function $F : \mathbb{R}^2 \to \mathbb{R}$ be given by

$$F(x, y) = \begin{cases} 1, & x = y^2, \ y > 0, \\ 0, & \text{otherwise} \end{cases}$$

is Gâteaux differentiable at $(0, 0)$ while not being continuous there.

Exercise 3.11

Prove that function $f : \mathbb{R}^2 \to \mathbb{R}$ given in polar coordinates

$$f(x, y) = r \cos 3\varphi, \ (x, y) = (r \cos \varphi, r \sin \varphi)$$

has the Gâteaux variation at $(0, 0)$, but not the Gâteaux derivative there.

Exercise 3.12

Prove that the functional considered in Example 3.7 is Gâteaux differentiable.

For Gâteaux derivatives one has the well known *Fermat Rule* which follows directly as in real line case:

Theorem 3.3 (Fermat Rule)

Let E be a real Banach space and let $F : E \to \mathbb{R}$. If $x_0 \in E$ is a minimizer of F over E and if F is differentiable in the sense of Gâteaux (at least at x_0), then for each $h \in E$ we have $\left\langle F'(x_0), h \right\rangle = 0$.

Proof Since F is differentiable in the sense of Gâteaux at x_0, it has the Gâteaux variation which implies by Theorem 3.2 that

$$\left\langle F'(x_0), h \right\rangle \geq 0.$$

for any direction $h \in E$. Taking $-h$ in the above we obtain that

$$\left\langle F'(x_0), h \right\rangle \leq 0$$

which implies the assertion.

Using the aforementioned result together with *the du Bois-Reymond Lemma* we now derive necessary optimality conditions for the basic problem of the calculus of variations. Other problems in the calculus of variation are treated likewise, but require a lot of background study for which we direct the Reader to [49] and next to [4], as well as to [23].

Example 3.14 (Basic Problem of the Calculus of Variations)

In the space $E = C_0^1[0, 1]$ we look for minimizers of the following action functional

$$F(u) = \int_0^1 f\left(t, u(t), u'(t)\right) dt = \int_0^1 f[u(t)]dt,$$

under assumption that f, f_u, $f_z \in C\left([0, 1] \times \mathbb{R}^2\right)$, where by f_u and f_z we understand the partial derivatives of f with respect to the second and the third variable. We recall that this functional has already been investigated in Example 3.7. We have derived the formula for Gâteaux variation at any point $u \in E$ in each direction $h \in E$

$$\delta F(u; h) = \int_0^1 \left(f_u[u(t)]h(t) + f_z[u(t)]h'(t)\right) dt$$

which we can easily prove to be the Gâteaux derivative in fact. Then if $u_0 \in E$ is a minimizer of F we obtain that

$$\int_0^1 \left(f_u[u_0(t)]h(t) + f_z[u_0(t)]h'(t)\right) dt = 0$$

which by the suitable version of *the du Bois–Reymond Lemma*, see Corollary 2.2, leads to the following equation

$$\frac{d}{dt} f_u\left(t, u_0(t), u_0'(t)\right) = f_z\left(t, u_0(t), u_0'(t)\right) \text{ for } t \in [0, 1]$$

called *the Euler–Lagrange equation* for the functional F, i.e., an equation satisfied by any critical point.

Exercise 3.13

Derive the Euler–Lagrange equation for the functional F from Example 3.14 in case it does not depend either on u or on u'.

Exercise 3.14

Let $E = C_0^1 [0, 1]$. Assume that both the function $g : [0, 1] \times \mathbb{R} \to \mathbb{R}$ and its derivative with respect to the second variable are jointly continuous over $[0, 1] \times \mathbb{R}$. Consider $F : E \to \mathbb{R}$, given by formula (3.3), i.e.

$$F(u) = \int_0^1 g(t, u(t)) \, dt + \frac{1}{2} \int_0^1 |u'(t)|^2 \, dt.$$

Prove that any minimizer u of F satisfies the following the Euler–Lagrange equation:

$$u''(t) + g_u(t, u(t)) = 0.$$

Derive the Euler–Lagrange equation in case $E = C^1 [0, 1]$.

Now we show how to use the Euler–Lagrange equation for finding minimizers of action functionals (provide that we know that these exist).

Example 3.15

We will determine a curve $u \in C^1[0, 1]$ with smallest length which connects the two end points $(0, x_1)$ and $(1, x_2)$ (where $x_1, x_2 \in \mathbb{R}$). It is easy to note that such a curve is a straight line connecting both points. Nevertheless, in order to prove this conjecture, we minimize

$$F(u) = \int_0^1 \sqrt{1 + (u'(t))^2} \, dt$$

over the following convex set

$$S = \left\{ u \in C^1[0, 1] \mid u(0) = x_1 \text{ and } u(1) = x_2 \right\}.$$

Supposing this problem is solvable, we see that the Euler–Lagrange equation now reads

$$\frac{d}{dt} \frac{2u'(t)}{2\sqrt{1 + u'(t)}} = 0$$

Then we get for some constant $c \in \mathbb{R}$

$$\frac{u'(t)}{\sqrt{1 + (u'(t))^2}} = c \text{ for all } t \in [0, 1]$$

and finally we obtain

$$u' \equiv \text{constant}$$

which leads directly to the assertion when we take into account that $u(t) = at + b$ and when we use the fact that $u \in S$.

Exercise 3.15

Consider functional

$$F(u) = \int_0^1 \cos u(t) dt$$

on $E = C[0, 1]$ and next on $E = \{u \in C[0, 1] : u(0) = u(1) = \pi\}$. Formulate and solve the relevant Euler–Lagrange equation.

Exercise 3.16

Consider functional

$$F(u) = \int_0^1 \left[(u(t))^3 + e^t u(t) \right] dt$$

on the space $E = C[0, 1]$ and next on $E = \{u \in C[0, 1] : u(0) = u(1) = 0\}$. Formulate and solve the relevant Euler–Lagrange equation.

3.4 The Fréchet Derivative

Now we turn to the notion of the Fréchet derivative, which is the strongest of all already introduced and which implies the continuity.

Definition 3.5 (Fréchet Derivative)

Let E be a real Banach space. Functional $F : E \to \mathbb{R}$ is Fréchet differentiable at $x_0 \in E$ if there is a linear continuous functional $F'(x_0) \in E^*$ such that

$$\lim_{\|h\| \to 0} \frac{\left| F(x_0 + h) - F(x_0) - \langle F'(x_0), h \rangle \right|}{\|h\|} = 0. \tag{3.9}$$

Note that if we introduce an equivalent norm on E then we do not alter the Fréchet differentiability. In compliance with what was observed about the Gâteaux derivative, we see that (3.9) can be written as

$$F\left(x+h\right) - F\left(x\right) - \langle F'(x), h\rangle = o\left(h\right),$$

where $o\left(h\right)$ is the Landau symbol understood now as follows:

$$\lim_{\|h\| \to 0} \frac{o\left(h\right)}{\|h\|} = 0. \tag{3.10}$$

In contrast to the definition of the Gâteaux derivative the limit in (3.10) is taken over all admissible directions and not on each direction separately. Higher-order derivatives are also being introduced and these correspond to multilinear maps. We will not follow this direction retaining for higher-order conditions the n−th Gâteaux variation, which suffice as far as our applications are concerned.

Exercise 3.17

Let E be a real Banach space. Assume that the functional $F : E \to \mathbb{R}$ is differentiable in the sense of Fréchet at some $x \in E$. Show that it is differentiable in the sense of Gâteaux and has the Gâteaux derivative and the Gâteaux variation at this point which all equal.

Another difference with respect to the Gâteaux derivative is contained in the following lemma, which provides the required relation between differentiability and continuity, again it reflects what we know about the weak and the strong differentiability in $\mathbb{R}^N$:

Lemma 3.3

Let E be a real Banach space. Assume that the functional $F : E \to \mathbb{R}$ is differentiable in the sense of Fréchet at some $x \in E$. Then it is continuous at least at $x \in E$.

Proof Indeed, from (3.9) it follows that for some fixed $\varepsilon > 0$, there is $\delta > 0$ such that

$$\|F(x+h) - F(x)\| - \|\langle F'(x), h\rangle\| \le \|F(x+h) - F(x) - \langle F'(x), h\rangle\| \le \varepsilon\|h\|$$

for all $\|h\| < \delta$, $h \in E$. The above implies immediately that

$$\|F(x+h) - F(x)\| \le \left(\varepsilon + \|F'(x)\|\right)\|h\|.$$

Hence letting $\|h\| \to 0$ we see that F is continuous at x.

Checking the Fréchet is not as immediate as checking the Gâteaux differentiability. We now follow with some standard example:

Example 3.16

Consider the continuous function $F : \mathbb{R}^2 \to \mathbb{R}$

$$F(x, y) = \begin{cases} \frac{x^2 y}{x^2 + y^2} & \text{if } (x, y) \neq (0, 0), \\ 0 & \text{if } (x, y) = (0, 0) \end{cases}$$

which is identically zero on both coordinate axes and with both partial derivatives equal 0 at $(0, 0)$. Calculating (3.10) we have to check whether

$$\lim_{(h_1, h_2) \to (0,0)} \frac{(h_1)^2 h_2}{\left((h_1)^2 + (h_2)^2\right)^{3/2}} = 0.$$

Taking $h_1 = h_2 = t$ we have

$$\lim_{t \to 0^\pm} \frac{t^3}{2\sqrt{2}t^2|t|} = \pm\frac{1}{2\sqrt{2}} \neq 0.$$

Thus, the function F is not differentiable in the sense of Fréchet but it has the Gâteaux variation in each direction.

Exercise 3.18

Check if the function f from Example 3.16 if differentiable in the sense of Gâteaux.

Exercise 3.19

Check the differentiability of the following function

$$F(x, y, z) = \begin{cases} \left(x^2 + y^2 + z^2\right) \sin \frac{1}{\sqrt{x^2 + y^2 + z^2}} & \text{if } (x, y, z) \neq (0, 0), \\ 0 & \text{if } (x, y, z) = (0, 0). \end{cases}$$

Exercise 3.20

Find values α for which the function

$$F(x, y) = \begin{cases} |y|^\alpha \sin x & \text{if } y \neq 0, \\ 0 & \text{if } y = 0 \end{cases}$$

is Gâteaux and Fréchet differentiable.

When we look for the Gâteaux variation and prove that it defines a linear and continuos functional, we are done with having the Gâteaux derivative. For the Fréchet differentiability we have to prove that relation (3.10) holds, which is not an easy task in general. However there is some sufficient condition related to the Gâteaux differentiability which implies the Fréchet differentiability and which does not require checking condition (3.10) directly.

Definition 3.6

Let E be a real Banach space. Functional $F : E \to \mathbb{R}$ is continuously Gâteaux differentiable at $x_0 \in E$, if it is Gâteaux differentiable at least over some neighborhood of x_0 and if the mapping $F' : E \to E^*$ is continuous at x_0.

We consider continuity of the derivative over the whole space in a usual manner.

Lemma 3.4

Let E be a real Banach space. Assume that functional $F : E \to \mathbb{R}$ is continuously Gâteaux differentiable over E. Then it is continuously Fréchet differentiable over E and both derivatives coincide.

We will direct the reader about the proof of the above lemma after introducing *the Mean Value Theorem*. We want to use this tool immediately, and that is why we postpone its proof until some time later. If a functional is continuously Fréchet differentiable over E, then it is called a C^1 functional. The continuity of F' at x_0 is understood as usual: $x_n \to x_0$ in E implies $F'(x_n) \to F'(x_0)$ in E^*. Since the direct usage of the norm in E^* (the dual norm) is not always easy, so we recall that we can understand the continuity of F' as follows:

$$\langle F'(x_n), h \rangle \mapsto \langle F'(x_0), h \rangle \text{ as } n \to +\infty$$

uniformly in h taken from the unit sphere in E.

Example 3.17

Recalling Example 3.12 we can now prove that functional

$$x \mapsto \frac{1}{2}(x, x)_E = \frac{1}{2}\|x\|^2,$$

whose Gâteaux derivative at any fixed point x reads

$$h \to (x, h)_E,$$

is in fact C^1. Indeed taking $(x_n) \subset E$ such that $x_n \to x_0$ we see by the Schwarz Inequality for $\|h\| = 1$

$$|(x_n - x_0, h)_E| \leq \|x_n - x_0\| \, \|h\| \leq \|x_n - x_0\| \to 0.$$

Example 3.18

We recall that in Example 3.13 we assumed that $g : [0, 1] \times \mathbb{R} \to \mathbb{R}$ is an L^1−Carathéodory function such that the partial derivative $g_u : [0, 1] \times \mathbb{R} \to \mathbb{R}$ of g with respect to the second variable exists and is an L^1−Carathéodory function as well. We proved that functional $F : H^1_0(0, 1) \to \mathbb{R}$ given by formula

$$F(u) = \int_0^1 g\,(t, u(t))\,\mathrm{d}t + \frac{1}{2} \int_0^1 |\dot{u}(t)|^2 \,\mathrm{d}t$$

is Gâteaux differentiable over $H^1_0(0, 1)$ with the Gâteaux derivative given by

$$\langle F'(x), h \rangle = \int_0^1 g_u\,(t, u(t))\, h\,(t)\,\mathrm{d}t + \int_0^1 \dot{u}(t)\dot{h}(t)\mathrm{d}t$$

for all $u, h \in H^1_0(0, 1)$. Using technique applied in Example 2.8 it is immediate to note that F is a C^1 functional (if we also take into account Example 3.17).

Exercise 3.21

Let E be a real Hilbert space. Prove that the functional $F : E \to \mathbb{R}$ defined by

$$F(x) = \sqrt{(x, x)_E} = \|x\|$$

is differentiable in the sense of Fréchet at all $x \neq 0_E$ with a derivative defined by

$$h \mapsto \frac{1}{\sqrt{(x, x)_E}}\,(x, h)_E \quad \text{for all } x \neq 0_E \text{ and all } h \in E.$$

Exercise 3.22

Let $p \geq 2$. Let E be a real Hilbert space. Differentiate the following functional

$$F(x) = \frac{1}{p}\,\|x\|^p\,.$$

Prove that

$$\langle F'(x), h \rangle = \|x\|^{p-2}\,(x, h)_E = \|x\|^{p-2}\,(x, h)_E \quad \text{for all } x, h \in E.$$

Exercise 3.23

Prove that functional $f : \mathbb{R} \to \mathbb{R}$ defined by

$$F(x) = \begin{cases} x^2, & x \text{ rational}, \\ 0, & x \text{ otherwise} \end{cases}$$

is differentiable in the sense of Fréchet at $x = 0$ but it is not continuously differentiable in the sense of Fréchet there.

Exercise 3.24

Let A be a $N \times N-$matrix. Find the Fréchet derivative of $F : \mathbb{R}^N \to \mathbb{R}$ given by

$$F(x) = \frac{1}{2} |Ax|^2.$$

Exercise 3.25

Let U be an open set of $\mathbb{R}^N$ and let $u, v : U \to \mathbb{R}^m$ be Fréchet differentiable. Define $F : U \to \mathbb{R}$ by

$$F(x) = u(x) \cdot v(x).$$

Prove that F is Fréchet differentiable and that the derivative of F satisfies for all $x \in \mathbb{R}^N$

$$F'(x) = u(x) \cdot v'(x) + v(x) \cdot u'(x).$$

Exercise 3.26

Let E be a Hilbert space and let $A : E \to E$ be a linear continuous and self-adjoint operator. Find the Fréchet derivative of $F : E \to \mathbb{R}$ given by

$$F(x) = \frac{1}{2} (Ax, x)_E.$$

Exercise 3.27

Find points at the which the following functional $F : \mathbb{R}^N \to \mathbb{R}$ is not differentiable in the sense of Fréchet when:

(a) $F(x) = \max_{1 \le i \le N} |x_i|$;
(b) $F(x) = \sum_{i=1}^{N} |x_i|$.

Exercise 3.28

Let $E = C[0, 1]$. Find the Fréchet derivative of $F : E \to \mathbb{R}$ when

(a) $F(x) = x(0)$;
(b) $F(x) = x^2(1)$;
(c) $F(x) = \sin x(1)$.

Check what happens when $C[0, 1]$ is replaced with $H_0^1(0, 1)$?

3.5   On the Differentiability of Maps Between Banach Spaces

For the Chain Rule, which we are going to introduce, we need a definition of a derivative for maps between normed spaces, which is as follows. Let X and Y be real Banach spaces; $\mathcal{L}(X, Y)$ stands for the space consisting of continuous linear mappings from X into Y. A mapping $f : X \to Y$ is said to be Gâteaux differentiable at $x_0 \in X$ if there exists $f'(x_0) \in \mathcal{L}(X, Y)$ such that for every $h \in X$

$$\lim_{t \to 0} \frac{f(x_0 + th) - f(x_0)}{t} = f'(x_0)h.$$

Note that when Y is different from $\mathbb{R}$, it is evident that the derivative at x_0 is no longer a functional. This is why we introduce a different symbol for it. The operator $f'(x_0)$ is then called, as was the case with functionals, the Gâteaux derivative of f at x_0. A mapping f is continuously Gâteaux-differentiable if

$$f' : X \ni x \mapsto f'(x) \in \mathcal{L}(X, Y)$$

is continuous in the relevant topologies. Note that the introduction of equivalent norms on X and Y does not influence the Gâteaux differentiability of a mapping. An operator $f : X \to Y$ is said to be Fréchet-differentiable at $x_0 \in X$ if there exists a continuous linear operator $f'(x_0) \in \mathcal{L}(X, Y)$ such that

$$\lim_{\|h\| \to 0} \frac{\|f(x_0 + h) - f(x_0) - f'(x_0)h\|}{\|h\|} = 0.$$

The operator $f'(x_0)$ is called the Fréchet derivative of f at x_0. When f is Fréchet differentiable it is also continuous and Gâteaux differentiable and obviously both derivatives coincide. A mapping f is continuously Fréchet differentiable if

$$f' : X \ni x \mapsto f'(x) \in \mathcal{L}(X, Y)$$

is continuous. If f is continuously Gâteaux differentiable then it is also continuously Fréchet differentiable. The space of all continuously Fréchet differentiable mappings from X into Y will be denoted by $\mathbf{C}^1(X, Y)$ or simply $\mathbf{C}^1$ if there is no ambiguity about the spaces.

We see also that we can define the Fréchet differentiability at x_0 requiring that

$$f(x_0 + h) - f(x_0) - f'(x_0)h = o\left(\|h\|\right)$$

or more explicitly that

$$f(x_0 + h) - f(x_0) - f'(x_0)h = r\left(\|h\|\right)\|h\|, \tag{3.11}$$

where $r(t) \to 0$ as $t \to 0^+$. We will use formula (3.11) for the sake of *the Chain Rule* further on.

Exercise 3.29

Assume that $f, g : X \to Y$ are differentiable (in either sense) at $x \in X$. Prove that for any $a, b \in \mathbb{R}$ the linear combination $af + bg$ is differentiable at x and the derivative obeys the following law

$$(af + bg)'(x) = af'(x) + bg'(x).$$

Exercise 3.30

Assume that $f : X \to Y$ is differentiable (in either sense) at $x \in X$ and that $g : X \to \mathbb{R}$ is differentiable in the same sense. Prove that their product is also differentiable and find the formula for the derivative.

Example 3.19

Let $E = C[0, 1]$. Let $h : \mathbb{R} \to \mathbb{R}$ be continuously differentiable. Consider the mapping

$$H : E \to E$$

defined by the relation

$$H(u)(\cdot) = h(u(\cdot)).$$

We shall show that the mapping H is Fréchet differentiable at any point u_0 and we shall evaluate its derivative. Since we have

$$h(t_0 + t) = h(t_0) + h'(t_0)t + o(|t|) \text{ for any } t_0 \in \mathbb{R}$$

it immediately follows that

$$h(u_0(t) + h(t)) = h(u_0(t)) + h'(u_0(t))h(t) + o(|h(t)|) \text{ for } t \in [0, 1]$$

for any fixed function $u_0 \in E$ and for any direction $v \in E$. Since

$$|v(t)| \leqslant \|v\|_\infty \text{ for all } t \in [0, 1] \,,$$

we have that

$$h\left(u_0(t) + v(t)\right) = h\left(u_0(t)\right) + h'\left(u_0(t)\right) v(t) + o(\|v\|_\infty) \text{ for } t \in [0, 1] \,.$$

But this means that the mapping H is Fréchet differentiable at u_0, and its derivative reads:

$$H'\left(u_0(t)\right) v(t) = h'\left(u_0(t)\right) v(t) \text{ for } t \in [0, 1] \,.$$

We conclude this section with *the Chain Rule* and *the Mean Value Theorem*, which are given in order to warn the reader that not all intuitions from the real line can be now directly extended.

Example 3.20

Let $X = Y = \mathbb{R}^2$, $Z = \mathbb{R}$ and let $\varphi : \mathbb{R}^2 \to \mathbb{R}^2$ be a C^1 mapping defined by

$$\varphi\left(x, y\right) = \left(x^2, y\right) .$$

Let $f : \mathbb{R}^2 \to \mathbb{R}$ be a Gâteaux differentiable mapping given by

$$f\left(x, y\right) = \begin{cases} 1, & x = y^2, y > 0, \\ 0, & \text{otherwise} \end{cases}$$

which is Gâteaux differentiable at $(0, 0)$ and not continuous at $(0, 0)$. Then mapping $g = f \circ \varphi$ has the following form

$$g\left(x, y\right) = \begin{cases} 1, & |x| = y > 0, \\ 0, & \text{otherwise.} \end{cases}$$

We see that g in contrast to f is not differentiable in the sense of Gâteaux.

Example 3.21

Consider the mapping $f : \mathbb{R} \to \mathbb{R}^2$ defined by

$$f\left(t\right) = \left(\sin t, \cos t\right) .$$

While $f\left(2\pi\right) - f\left(0\right) = (0, 0)$, we easily see that there is no $c \in (0, 2\pi)$ such that $f'\left(c\right) = (0, 0)$.

Theorem 3.4 (Chain Rule)

Let X, Y, Z be real Banach spaces and assume that mappings $f : Y \to Z$, $g : X \to Y$ are differentiable in the sense of Fréchet. Then mapping $f \circ g$ is differentiable differentiable in the sense of Fréchet as well. Moreover,

$$(f \circ g)'(x) = f'(g(x)) g'(x) \text{ for } x \in X.$$

Proof We use formula (3.11). We put $y = g(x)$. It follows from the Fréchet differentiability of f that

$$f(y + v) = f(y) + f'(y)v + \|v\| r_1 (\|v\|). \tag{3.12}$$

Now we choose v as follows

$$v = g(x + h) - g(x) = g'(x)h + \|h\| r_2 (\|h\|)$$

and we further obtain from (3.12)

$$
\begin{aligned}
f(g(x + h)) = {} & f(y) + f'(y)g'(x)h \\
& + \|g(x + h) - g(x)\| r_1 (\|g(x + h) - g(x)\|) \\
& + f'(y) \|h\| r_2 (\|h\|).
\end{aligned}
$$

We see that the remainder

$$r(\|h\|) = \frac{\|g(x + h) - g(x)\| r_1 (\|g(x + h) - g(x)\|)}{\|h\|} + f'(y) r_2 (\|h\|)$$

has the property required for the Fréchet differentiability. Indeed, we have the following estimation

$$
\begin{aligned}
\|g(x + h) - g(x)\| r_1 (\|g(x + h) - g(x)\|) = {} & \|g'(x)h + \\
\|h\| r_2 (\|h\|) \| r_1 (\|g(x + h) - g(x)\|) \leq {} & \\
\big(\|g'(x)\| \|h\| + \|h\| r_2 (\|h\|)\big) r_1 (\|g(x + h) - g(x)\|). &
\end{aligned}
$$

Then, invoking the continuity of g, we see that $r(\|h\|) \to 0$ as $\|h\| \to 0$.

Exercise 3.31

Check whether the Chain Rule holds in case the inner mapping has the Gâteaux variation at any point. What sort of differentiability is then obtained?

In case the outer mapping, i.e., f in the aforementioned formalism, is not differentiable in the sense of Fréchet, we may not apply *the Chain Rule* as seen by Example 3.20.

Theorem 3.5 (The Mean Value Theorem)
Let E be a real Banach space. Let a functional $F : E \to \mathbb{R}$ be continuously differentiable. Then for all $x, h \in E$

$$\text{(a)}\ F(x + h) - F(x) = \int_0^1 \langle F'(x + th), h \rangle \, dt,$$

$$\text{(b)}\ |F(x + h) - F(x)| \leqslant \sup_{0 \leqslant t \leqslant 1} \left\| F'(x + th) \right\|_{E^*} \cdot \|h\|_E.$$

Proof Fix $x, h \in E$ and set $\varphi(t) = F(x + th)$, $\varphi : [0, 1] \to \mathbb{R}$. Then we have

$$\frac{d\varphi(t)}{dt} = \langle F'(x + th), h \rangle$$

for all $t \in [0, 1]$. The assertion (a) follows now from the classical Newton-Leibnitz formula applied to φ on $[0, 1]$. Using estimation

$$\left| \int_0^1 F'(x + th) h \, dt \right| \leqslant \sup_{0 \leqslant t \leqslant 1} \left\| F'(x + th) \right\|_{E^*} \cdot \|h\|_E$$

we see that (b) is a simple consequence of (a).

▶ **Remark 3.7** In the aforementioned, we may assume that U is an open subset of E, and that a functional $F : U \to \mathbb{R}$ is Gateaux differentiable at every point of the interval $[x, x + h] \subset U$ with the derivative being continuous there.

Exercise 3.32

Prove the following version of *the Mean Value Theorem:* Let $F : \mathbb{R}^N \to \mathbb{R}$ be a differentiable mapping. Then for any distinct $a, b \in \mathbb{R}^N$ there exists an $\overline{x} \in [a, b]$ different from a, b such that

$$F(b) - F(a) = \langle \nabla F(\overline{x}), b - a \rangle.$$

Hint: Put $u(t) = a + t(b - a)$ for $t \in [0, 1]$ and apply the classical Lagrange Mean Value to the function $t \mapsto F(u(t))$.

Exercise 3.33

Check if the aforementioned exercise can be generalized to functionals defined on a Banach space and differentiable in some suitable sense (which?).

Exercise 3.34

Under the assumptions of Theorem 3.5 prove that for any $\Lambda \in E^*$

$$|F(x+h) - F(x) - \langle \Lambda, h \rangle| \leq \sup_{0 \leq t \leq 1} \left\| F'(x+th) - \Lambda \right\|_{X^*} \cdot \|h\|_E \text{ for all } x, h \in E$$

Argue that

$$\left| F(x+h) - F(x) - \left\langle F'(z), h \right\rangle \right| \leq \sup_{0 \leq i \leq 1} \left\| F'(x+th) - F'(z) \right\|_{E^*} \cdot \|h\|_E$$

(3.13)

for fixed $x, h \in X$ with $z \in (x, x+h)$.

Exercise 3.35

Prove Lemma 3.4. Hint: Use estimation (3.13).

Exercise 3.36

Let E be a real Banach space. Prove that if a functional $F : E \to \mathbb{R}$ is continuously differentiable, then it is locally Lipschitz continuous.

3.6   More on the Convexity

We follow with some properties of convex functionals with reference to the differentiability.

Theorem 3.6 (Convexity Criteria)
Let E be a real Banach space. Let $F : E \to \mathbb{R}$ be Gâteaux differentiable. Then F is convex if and only if

$$F(x) - F(y) \geq \left\langle F'(y), x - y \right\rangle \text{ for all } x, y \in E. \tag{3.14}$$

Proof Let F be a Gâteaux differentiable convex functional and let $x, y \in E$, $\lambda \in (0, 1]$. We have

$$F((1-\lambda)x + \lambda y) \leq (1-\lambda)F(x) + \lambda F(y)$$

which implies that

$$\frac{F((1-\lambda)x + \lambda y)' - F(x)}{\lambda} < F(y) - F(x).$$

Now we easily calculate that

$$\langle F'(x), (y-x)\rangle = \lim_{\lambda \to 0^+} \frac{F((1-\lambda)x + \lambda y) - F(x)}{\lambda} \leq F(y) - F(x).$$

Now let us assume that (3.14) hold. Take $x, y \in E$ and $\lambda \in [0, 1]$. For pairs

$$(x, \lambda x + (1-\lambda)y), \quad (y, \lambda x + (1-\lambda)y)$$

we have

$$F(x) \geq F(\lambda x + (1-\lambda)y) + (1-\lambda)\langle F'(\lambda x + (1-\lambda)y), (x-y)\rangle \quad / \cdot \lambda$$

$$F(y) \geq F(\lambda x + (1-\lambda)y) + \lambda\langle F'(\lambda x + (1-\lambda)y), (y-x)\rangle \quad / \cdot (1-\lambda)$$

Summing up both sides, we have

$$\lambda F(x) + (1-\lambda)F(y) \geq F(\lambda x + (1-\lambda)y).$$

Exercise 3.37

Check whether in the aforementioned theorem it suffices to use first Gâteaux variation instead of the Gâteaux derivative. What about the Lagrange variation?

▶ **Remark 3.8** Assume that $J : \mathbb{R} \to \mathbb{R}$ is continuous and convex. Then the functional

$$C[0, 1] \ni u \to \int_0^1 J(u(t))\,dt$$

is also convex. Indeed, since J is convex it follows for any $u, v \in C[0, 1]$ and all $\lambda \in [0, 1]$

$$J(\lambda u(t) + (1-\lambda)v(t)) \leq \lambda J(u(t)) + (1-\lambda)J(v(t)) \text{ for } t \in [0, 1].$$

Hence

$$\int_0^1 J(\lambda u(t) + (1-\lambda)v(t))\,dt \leq \lambda \int_0^1 J(u(t))\,dt + (1-\lambda)\int_0^1 J(v(t))\,dt.$$

Exercise 3.38

Consider functional

$$H_0^1(0, 1) \ni u \rightarrow \int_0^1 J(t, u(t)) \, dt \tag{3.15}$$

where $J : [0, 1] \times \mathbb{R} \rightarrow \mathbb{R}$ is an L^1–Carathéodory function such that for a.e. $t \in [0, 1]$ function $x \longmapsto J(t, x)$ is convex. Prove that functional (3.15) is also convex.

Theorem 3.7 (Convexity vs. Minimizers)
Let E be a real Banach space. Let $F : E \rightarrow \mathbb{R}$ be convex. Then the following hold:

(a) each critical point (under the assumption that F is Gâteaux differentiable) is a minimizer of F;
(b) the set of minimizers is convex;
(c) when F is strictly convex, then it has at most one minimizer;
(d) each local minimizer is a global one.

Proof

(a) Let x_0 be a critical point. By the convexity of F we immediately obtain

$$F(x) \geq F(x_0) + \left\langle F'(x_0), (x - x_0) \right\rangle = F(x_0) \text{ for any } x \in E.$$

This means that x_0 is a global minimizer.
(b) Assume there are two minimizers $x_1, x_2 \in E$ and let $\lambda \in (0, 1)$. Then we have

$$\min_{x \in E} F(x) \leq F(\lambda x_1 + (1 - \lambda)x_2) \leq \lambda F(x_1) + (1 - \lambda)F(x_2) = \min_{x \in E} F(x)$$

Therefore $\lambda x_1 + (1 - \lambda)x_2$ is also a minimizer for each fixed $\lambda \in [0, 1]$. Note that the assertion is obvious for $\lambda = 0$ and $\lambda = 1$. Therefore the set of minimizers is convex.
(c) Let $x_1, x_2 \in E$ be two distinct minimizers of F which is assumed to be strictly convex. Fix $\lambda \in (0, 1)$. Then

$$\min_{x \in E} F(x) \leq F(\lambda x_1 + (1 - \lambda)x_2) < \lambda F(x_1) + (1 - \lambda)F(x_2) = \min_{x \in E} F(x)$$

which is a contradiction.

(d) Assume that x_0 is a local minimizer, i.e.

$$F(x) \geq F(x_0) \quad \text{for } x \in \overline{B}(x_0, r)$$

where $\overline{B}(x_0, r)$ is some closed ball centred at x_0. Let us take any $x \in E \setminus \overline{B}(x_0, r)$. Put

$$\lambda = \frac{r}{\|x - x_0\|} < 1.$$

Note that $\lambda x + (1 - \lambda)x_0 \in \overline{B}(x_0, r)$. Indeed,

$$\|\lambda x + (1 - \lambda)(x - x_0) - x_0\| = \|\lambda(x - x_0)\| = r.$$

Thus for any $x \in E$ and the above fixed λ we have

$$F(x_0) \leq F(\lambda x + (1 - \lambda)x_0) \leq \lambda F(x) + (1 - \lambda)F(x_0).$$

This implies

$$\lambda F(x_0) \leq \lambda F(x)$$

and so $F(x_0) \leq F(x)$ for all $x \in E$.

Exercise 3.39

Check if we can weaken the differentiability assumption in (a) of Theorem 3.7.

Some remarks are in order concerning the above results:

▶ **Remark 3.9** Note that convexity itself does not provide the sufficient condition for the existence of a minimizer. The classical example is that of a strictly convex function $F(x) = e^x$. This means that the aforementioned theorem has a value while characterizing the set of solutions to equation $F'(x) = 0$ but it does not pertain to its solvability. On the other hand it says that each critical point is a minimizer, therefore if we find a critical point of a convex functional, then we know that it is necessarily a minimizer. Hence, solutions to the Euler–Lagrange equation to convex functionals are necessarily minimizers.

Now we turn to second-order conditions for which we will use the second order *Gâteaux variation*.

Theorem 3.8

Let E be a real Banach space. Let $x_0 \in E$ and let $U(x_0)$ be the open neighborhood of x_0. Assume that functional $F : U(x_0) \to \mathbb{R}$ has second order Gâteaux variation over $U(x_0)$ and that $F'(x_0; h) = 0$ for each $h \in E$. Let there exist a constant $c > 0$ such that

$$F^{(2)}(x_0; h) \geq c \|h\|^2 \text{ for all } h \in E$$

and that for any $\varepsilon > 0$ there is some $\eta(\varepsilon)$ such that

$$\left| F^{(2)}(x_0; h) - F^{(2)}(x; h) \right| \leq \varepsilon \|h\|^2$$

for all $x, h \in E$ with $\|x - x_0\| \leq \eta(\varepsilon)$. Then x_0 is a local minimizer of F over $U(x_0)$.

Proof Let $\varepsilon = \frac{c}{2}$ and chose $\eta(\varepsilon)$ so that

$$\left| F^{(2)}(x_0 + h; h) - F^{(2)}(x_0; h) \right| \leq \frac{c}{2} \|h\|^2$$

for all $\|h\| \leq \eta(\varepsilon)$. Let $h \in E$ with $\|h\| \leq \eta(\varepsilon)$ be fixed. We apply the Taylor formula (with the Lagrange remainder) to the auxiliary function $g(t) = F(x_0 + th)$ defined on a neighborhood of 0 or radius $\eta(\varepsilon)$. Since $g'(0) = \delta F(x_0; h) = 0$ it follows

$$F(x_0 + h) - F(x_0) = g(1) - g(0) = \frac{1}{2} F^{(2)}(x_0 + \theta h; h)$$

where $0 < \theta < \eta(\varepsilon)$. Therefore

$$F(x_0 + h) - F(x_0) = \frac{1}{2} F^{(2)}(x_0 + \theta h; h) \geq \frac{1}{2} \left(c - \frac{1}{2} c \right) \|h\|^2.$$

But since $\|h\| \leq \eta(\varepsilon)$ it means that x_0 is a local minimizer.

Theorem 3.9 (Second-Order Convexity Condition)

Let E be a real Banach space. Assume that $C \subset E$ is open and convex, $F : C \to \mathbb{R}$ has second-order Gâteaux variation over C and for all $x \in C$, $h \in E$ we have

(Continued)

> **Theorem 3.9 (continued)**
>
> $$F^{(2)}(x; h) \geq 0.$$
>
> *Then functional F is convex over C. If moreover*
>
> $$F^{(2)}(x; h) > 0$$
>
> *for all $h \neq 0$, then functional F is strictly convex over C.*

Exercise 3.40

Prove Theorem 3.9. Hint: Use the auxiliary function defined in the proof of Theorem 3.8 and the second-order convexity test for functions defined on the real line.

Exercise 3.41

Let H be a real Hilbert space. Using Theorem 3.9 show that the functional $F : H \to \mathbb{R}$ defined by

$$F(x) = \frac{1}{2} \|x\|^2$$

is strictly convex.

Exercise 3.42

Let H be a real Hilbert space. Is functional $F : H \to \mathbb{R}$ defined by

$$F(x) = \frac{1}{2} \|x\|^4$$

strictly convex. Can this result be reached by Theorem 3.9?

On the Weierstrass Theorem in Infinite-Dimensional Spaces

4

In the (infinite-dimensional) real space Banach space E one can formulate and prove (with same technique as with Theorem 1.1 and that is why we omit formal arguments) the following classical **Weierstrass Theorem:**

Assume that $F : E \to \mathbb{R}$ is lower semicontinuous and that $D \subset E$ is compact. Then there is some $x_0 \in D$ such that

$$F(x) \geq F(x_0) \text{ for all } x \in D.$$

The aforementioned result is of not much interest to us due to the fact that compact sets in infinite-dimensional reflexive Banach spaces have empty interiors, in particular the closed unit ball is not compact. Thus, we will have to resort to weak topologies due to the following background result:

Theorem 4.1 (Sequential Weak Compactness of a Unit Ball)
Let E be a real reflexive Banach space. The closed unit ball in E is sequentially weakly compact, i.e., each sequence from the ball contains a weakly convergent subsequence.

▶ **Remark 4.1** The aforementioned result is valid if a closed unit ball is replaced with any closed and bounded convex set, which in this case becomes sequentially weakly closed and bounded.

In order to prepare this chapter, we decided to use [20, 32, 37] for the exposition of variational methods and [28] for some approach toward the Dirichlet Laplacian.

© The Author(s), under exclusive license to Springer Nature Switzerland AG 2024   105
M. Galewski, *Basics of Nonlinear Optimization*, Compact Textbooks in
Mathematics, https://doi.org/10.1007/978-3-031-77160-6_4

4.1   Direct Variational Method

Let E denote a real reflexive Banach space. Recall that the coercivity of functional $F : E \to \mathbb{R}$ implies that for each $\alpha \in \mathbb{R}$ the Lebesgue level sets $F^\alpha = \{x \in E : F(x) \leq \alpha\}$ are bounded, while the convexity of F provides that these are convex as well. Closedness of F^α on the other hand is guaranteed by the lower semicontinuity of F.

Theorem 4.2

Let E be a real reflexive Banach space. Let $F : E \to \mathbb{R}$ be sequentially weakly lower semicontinuous and let $D \subset E$ be sequentially weakly compact. Then F has at least one minimizer over D.

Proof Suppose F is not bounded from below on D. Then there is a sequence $(x_n) \subset D$ such that $\lim_{n \to +\infty} F(x_n) = -\infty$. Since E is reflexive, we see that (x_n) has a weakly convergent subsequence (x_{n_k}) with a weak limit $\tilde{x}$. Hence $\lim_{k \to +\infty} F(x_{n_k}) = -\infty$ and since F is sequentially weakly lower semicontinuous we see that

$$-\infty = \liminf_{k \to +\infty} F(x_{n_k}) \geq F(\tilde{x}),$$

which is impossible. Hence, F is bounded from below on D and it has a weakly convergent minimizing sequence $(x_n) \subset D$ with a limit x_0. We then have

$$\inf_{x \in D} F(x) = \liminf_{n \to +\infty} F(x_n) \geq F(x_0) \geq \inf_{x \in D} F(x).$$

Then x_0 is a minimizer of functional F over D.

In case $D = E$, the boundedness of any minimizing sequence is guaranteed when sets F^α for each $\alpha \in \mathbb{R}$ are bounded. This leads to the following version of the Weierstrass Theorem, which is used to derive the so-called *direct method of the calculus of variations:*

Theorem 4.3

Let E be a real reflexive Banach space. Let the functional $F : E \to \mathbb{R}$ be Gâteaux differentiable, sequentially weakly lower semicontinuous and coercive. Then F has at least one minimizer x_0 over E

$$F(x_0) = \inf_{x \in E} F(x)$$

(Continued)

Theorem 4.3 (continued)
which is also a critical point, that is

$$\langle F'(x_0), h \rangle = 0 \text{ for all } h \in E. \tag{4.1}$$

Proof Take such $\alpha \in \mathbb{R}$ that F^α is non-empty. Since functional F is coercive it follows that F^α is bounded. Since F is sequentially weakly lower semicontinuous it follows that F^α is additionally sequentially weakly closed. Hence F^α is sequentially weakly compact. We see at once that

$$\inf_{x \in E} F(x) = \inf_{x \in F^\alpha} F(x)$$

Taking $D = F^\alpha$ we can apply Theorem 4.2 in order to find an element $x_0 \in F^\alpha$ such that $\inf_{x \in F^\alpha} F(x) = F(x_0)$. Hence x_0 is a minimizer of F over E. When we apply the Fermat Rule the proof is finished.

Equation (4.1) is often called *the Euler–Lagrange equation* for functional F. Note that the aforementioned theorem does not guarantee that the critical point which we obtain is nontrivial. Moreover, the minimizer is approximated by a weakly convergent minimizing sequence.

Exercise 4.1

Assume in addition to the assumptions of Theorem 4.3 that F is convex. Prove that the set of minimizers of F is convex.

Exercise 4.2

Assume in addition to the assumptions of Theorem 4.3 that F is strictly convex. Prove that F has exactly one minimizer.

Exercise 4.3

Assume in Theorem 4.3 that functional F has the Gâteaux variation instead of the Gâteaux derivative. Provide the corresponding assertion about the existence and the uniqueness.

Exercise 4.4

Let E be a Hilbert space. Let $b \in E$ and $c \in \mathbb{R}$. Prove that functional $F : E \to \mathbb{R}$ defined by

$$F(x) = \frac{1}{2} \|x\|^2 + (b, x)_E + c$$

has exactly one minimizer. When the minimizer is nontrivial?

We give the following example showing what may happen when the functional is not sequentially weakly lower semicontinuous:

Example 4.1

Let $r > 0$ (small) be fixed and consider the closed unit ball $\overline{B\,(0,\,1)} \subset l^2$. Denote by $\left\{\overline{B\,(x_n,\,r)}\right\}$ for $n \geq 1$, the infinite family of closed balls which do not intersect and which are contained in $\overline{B\,(0,\,1)}$. We define

$$F\,(x) = \begin{cases} \|x\| - 1, & \|x\| \geq 1, \\ \left(1 - \frac{1}{n}\right)(\|x - x_n\| - r), & x \in \overline{B\,(x_n,\,r)}, \\ 0, & \text{otherwise.} \end{cases}$$

Then F is bounded from below by $-r$, it is coercive but the infimal value is not reached.

4.2   Some Remarks on Approximation Problems

Example 4.2

Let E be a real reflexive Banach space. Assume that $D \subset E$ is closed and convex and take $\hat{x} \notin D$. We shall show that there is at least one $x_0 \in D$ such that

$$\inf_{x \in D} \|x - \hat{x}\| = \|x_0 - \hat{x}\|.$$

Indeed, let us define the functional $F : E \to \mathbb{R}$ by the formula

$$F\,(x) = \|x - \hat{x}\|.$$

Then F is continuous and convex over E and D is sequentially weakly closed. Fix any $y \in D$ and consider a convex and closed set

$$S = \{x \in D : F\,(x) \leq F\,(y)\}$$

which is *the Lebesgue level set* for functional F. Moreover, due to the coercivity of F, set S is bounded. Therefore, S is sequentially weakly compact. Since F is sequentially weakly lower semicontinuous, we reach the assertion by Theorem 4.2.

Exercise 4.5

Prove that in case E is a real Hilbert space, the point x_0 given earlier is unique.

With reference to Example 4.2 we have the following definition:

Definition 4.1 (Proximal set)

Let E be a real Banach space. Let $\varnothing \neq S \subset E$. The set S is called *proximal* if for every $\hat{x} \in E$ there is a point $\bar{x} \in S$ with the property

$$\|\bar{x} - \hat{x}\| \leq \|x - \hat{x}\| \text{ for all } x \in S. \tag{4.2}$$

In this case $\bar{x}$ is called best approximation to $\hat{x}$ from S.

Thus, from Example 4.2 we learn that:

Lemma 4.1

Let E be a real reflexive Banach space. Each closed and convex nonempty subset of E is proximal.

We have a well-known result about reflexive spaces:

Theorem 4.4 (James Theorem)

A real Banach space E is reflexive if and only if every continuous linear functional from E^* attains its supremum on the closed unit ball in E.

The James Theorem also implies the following sufficient condition about the weak sequential compactness of a nonempty subset of E (in case the reflexivity of E is not assumed since otherwise the assertion is trivial):

Corollary 4.1

Let S be a nonempty convex bounded closed subset of a real Banach space E. If every continuous linear functional attains its supremum on S then the set S is weakly sequentially compact.

Now we relate reflexivity to the notion introduced in the definition of proximal set:

Theorem 4.5

A real Banach space E is reflexive if and only if every nonempty convex closed subset is proximal.

Proof By Lemma 4.1 we need to prove that if each nonempty convex closed subset is proximal, then E is reflexive. Assume that E is not reflexive. Then the closed unit ball

$$\overline{B}\,(0_E,\, 1) := \{x \in E \mid \|x\| \leq 1\}$$

is not weakly sequentially compact and by the James Theorem there is a continuous linear functional l which does not attain its supremum on the set $\overline{B}\,(0_E,\, 1)$, i.e.,

$$l(x) < \sup_{y \in \overline{B}(0_E,1)} l(y) \quad \text{for all } x \in \overline{B}\,(0_E,\, 1).$$

Let us define the convex closed set

$$S := \left\{ x \in E \mid l(x) \geq \sup_{y \in \overline{B}(0_E,1)} l(y) \right\}.$$

Then one obtains $S \cap \overline{B}\,(0_E,\, 1) = \emptyset$. Consequently the set S is not proximal.

Example 4.3

Now let S be a nonempty subset of a non-reflexive spaces $C\,[0,\, 1]$, and let $\hat{x} \in C\,[0,\, 1]$ be fixed. We are looking for a function $\bar{x} \in S$ satisfying the property (4.2). Due to Lemma 4.1 we rather should expect that we cannot get rid of the type of reflexivity. Thus, we solve the aforementioned posed problem if S is a nonempty convex closed subset of the normed space $C\,[0,\, 1]$ such that for any $\tilde{x} \in S$ the linear subspace spanned by $S - \{\tilde{x}\}$ is reflexive.

Exercise 4.6

Let S be a nonempty convex closed subset of the normed space $C\,[0,\, 1]$ such that for any $\tilde{x} \in S$ the linear subspace spanned by $S - \{\tilde{x}\}$ is reflexive. Prove that S is proximal.

We conclude with a non-existence and a non-uniqueness examples:

Example 4.4 (Non-existence)

Let $E := \{f \in C[0,\, 1] \mid f(0) = 0\}$, which is a (non-reflexive) Banach space when considered with the supremum norm and let

$$A := \left\{ g \in E \mid \int_0^1 g(t)\mathrm{d}t = 0 \right\}.$$

Then A is a closed subspace of E. Let $f(t) := t$ for $t \in [0, 1]$. Then $f \in E \backslash A$, since $f(0) = 0$ but $\int_0^1 f(t)dt = 1/2 \neq 0$. Hence

$$\frac{1}{2} = \int_0^1 f(t) - g(t)dt \leq \int_0^1 |f(t) - g(t)|dt \leq \|f - g\|_\infty \text{ for every } g \in A,$$

but there is no $g \in A$ such that $\|f - g\|_\infty = \frac{1}{2}$.

Example 4.5 (Non-uniqueness)

Consider $E = \mathbb{R}^2$ with the norm $\|x\|_1 = |x_1| + |x_2|$ and A being the straight line through the points

$$(-1, 1), (0, 0), (1, -1).$$

If $x := (1, 1)$, then for every $(\lambda, -\lambda)$ with $\lambda \in [-1, 1]$ we obtain

$$\|(\lambda, -\lambda) - (1, 1)\|_1 = |1 - \lambda| + |1 + \lambda| = 2.$$

4.3   Minimization of the Classical Euler Action Functional

Now we give a number of examples concerning minimization of functionals over function spaces. We exploit two approaches: we use Theorem 4.3 in case the functional is considered over a reflexive Banach space and it is coercive and sequentially weakly lower semicontinuous, and also we will also exploit the fact that for a convex functional $F : E \to \mathbb{R}$ each critical point is a global minimizer. The second method is somehow in contrast to the Weierstrass Theorem. Indeed, by the Weierstrass Theorem we know that the minimizer exists and this implies that the relevant Euler–Lagrange equation is solvable. While in the second approach in order to find a minimizer we must know that the Euler–Lagrange equation is solvable. This mimics what one does in calculus courses for finding extrema.

We start this section with some direct application of Theorem 4.3 to functional

$$J : H_0^1(0, 1) \to \mathbb{R}$$

given by

$$J(u) = \frac{1}{2} \int_0^1 |\dot{u}(t)|^2 dt + \frac{1}{4} \int_0^1 |u(t)|^4 dt + \int_0^1 g(t) u(t) dt, \tag{4.3}$$

where $g \in L^2(0, 1)$ is fixed. We see at once that J is well defined, i.e. for any fixed $u \in H_0^1(0, 1)$ we have that $J(u)$ is finite.

Exercise 4.7

Show that functional J given by (4.3) is continuously differentiable and (strictly) convex.

Exercise 4.8

Check whether J given by (4.3) is continuously differentiable when considered on $C^1[0, 1]$.

Theorem 4.6

The functional J given by (4.3) has exactly one critical point which is a minimizer.

Proof Since J is convex and continuous by Theorem 2.3 it follows that J is sequentially weakly lower semicontinuous over $H_0^1(0, 1)$. Since J is strictly convex, it has at most one minimizer. Thus in order to apply Theorem 4.3 it suffice to show that J is coercive.

From the Schwarz Inequality and the Poincaré Inequality, we see that

$$\left| \int_0^1 g(t) u(t) \, dt \right| \leq \|u\|_{L^2} \|g\|_{L^2} \leq \frac{1}{\pi} \|u\|_{H_0^1} \|g\|_{L^2} \text{ for all } u \in H_0^1(0, 1)$$

and next also

$$\frac{1}{4} \int_0^1 u^4(t) \, dt + \int_0^1 u(t) g(t) \, dt \geq -\frac{1}{\pi} \|u\|_{H_0^1} \|g\|_{L^2} \text{ for all } u \in H_0^1(0, 1).$$

Thus,

$$J(u) \geq \frac{1}{2} \|u\|_{H_0^1}^2 - \frac{1}{\pi} \|u\|_{H_0^1} \|g\|_{L^2} \text{ for all } u \in H_0^1(0, 1).$$

Hence,

$$J(u) \to +\infty \text{ as } \|u\|_{H_0^1} \to +\infty.$$

Therefore, by Theorem 4.3 there exists exactly one $u_0 \in H_0^1(0, 1)$ such that

$$\left\langle J'(u_0), h \right\rangle = 0 \text{ for all } h \in H_0^1(0, 1). \tag{4.4}$$

▶ **Remark 4.2**   Relation 4.4) means that

$$\int_0^1 \dot{u}_0(t)\,\dot{h}(t)\,dt + \int_0^1 \left(u_0^3(t) + g(t) \right) h(t)\,dt = 0 \text{ for all } h \in H_0^1(0, 1).$$

Using *the du Bois-Reymond Lemma* we see that

$$\ddot{u}_0(t) = u_0^3(t) + g(t) \text{ for a.e. } t \in [0, 1].$$

Hence $u_0 \in H_0^1(0, 1) \cap H^2(0, 1)$ is the unique solution of the following differential equation:

$$\ddot{u}(t) = u^3(t) + g(t)$$

with boundary conditions

$$u(0) = u(1) = 0.$$

Exercise 4.9

Follow the lines of the above remark if $g \in L^p(0, 1)$ for some $p \geq 1$ and if $g \in C[0, 1]$. What can we say about the solution in this case?

Exercise 4.10

Consider the functional

$$J : H_0^1(0, 1) \to \mathbb{R}$$

given by

$$J(u) = \frac{1}{2} \int_0^1 |\dot{u}(t)|^2\,dt + \int_0^1 \sin(u(t))\,dt + \int_0^1 g(t)\,u(t)\,dt,$$

where $g \in L^2(0, 1)$ is fixed. Prove that it is coercive and sequentially weakly lower semicontinuous. Argue that there is at least one minimizer to J. Formulate the relevant Euler–Lagrange equation.

Exercise 4.11

Consider the functional

$$J : H_0^1(0, 1) \to \mathbb{R}$$

given by

$$J(u) = \frac{1}{2} \int_0^1 |\dot{u}(t)|^2\,dt + \frac{1}{3} \int_0^1 (u(t))^3\,dt + \int_0^1 g(t)\,u(t)\,dt,$$

where $g \in L^2(0, 1)$ is fixed. Prove that it is sequentially weakly lower semicontinuous but it is not coercive. Compare the result obtained with the case of the following functional

$$J(x) = \frac{1}{2} \int_0^1 |\dot{u}(t)|^2 \, dt + \frac{1}{3} \int_0^1 |u(t)|^3 \, dt + \int_0^1 g(t) u(t) \, dt.$$

Exercise 4.12

Let $G : \mathbb{R} \to \mathbb{R}$ be a continuously differentiable function, which is bounded from below and consider a functional

$$J : H_0^1(0, 1) \to \mathbb{R}$$

given by

$$J(u) = \frac{1}{2} \int_0^1 |\dot{u}(t)|^2 \, dt + \int_0^1 G(u(t)) \, dt.$$

Prove that it is coercive and sequentially weakly lower semicontinuous. Argue that there is at least one minimizer to J. Formulate the relevant Euler–Lagrange equation and investigate the regularity of its solutions.

Now we turn to some special case of the aforementioned exercise:

Exercise 4.13

Let $G : \mathbb{R} \to \mathbb{R}$ be a continuously differentiable convex function with a derivative $g : \mathbb{R} \to \mathbb{R}$ and consider a functional

$$J : H_0^1(0, 1) \to \mathbb{R}$$

given by

$$J(u) = \frac{1}{2} \int_0^1 |\dot{u}(t)|^2 \, dt + \int_0^1 G(u(t)) \, dt. \tag{4.5}$$

Prove that J is continuously differentiable, sequentially weakly lower semi-continuous and coercive. Argue that J has exactly one minimizer. Hint: While proving the coercivity show that the following estimation holds

$$G(x) \geq -|g(0)| \, |x| + G(0) \quad \text{for all } x \in \mathbb{R}.$$

Find the Dirichlet problem, which is satisfied by the minimizer to functional J given by (4.5) and investigate the regularity of its solutions.

Example 4.6

In Example 3.3 we considered functional $F : C[0, 1] \to \mathbb{R}$ defined by

$$F(u) = \int_0^1 \left[\sin^3 t + u^2(t) \right] dt$$

and calculated for any $u, h \in C[0, 1]$ that

$$\left\langle F'(u), h \right\rangle = 2 \int_0^1 u(t)h(t)dt \text{ for } u, h \in C[0, 1].$$

The aforementioned formula says that F is C^1. We easily calculate that F is in fact strictly convex (by the second order test). The solution to equation

$$F'(u) = 0$$

when written as the Euler–Lagrange equation is of course $u_0 = 0$, which minimizes the given functional due to its convexity.

Example 4.7

Consider the functional $F : H_0^1(0, 1) \to \mathbb{R}$ defined by

$$F(u) = \int_0^1 (\dot{u}(t) - 1)^2 dt. \tag{4.6}$$

We easily calculate that F is strictly convex, coercive, and continuous. Indeed, the strict convexity of F follows from the strict convexity of the integrand. The continuity follows by a direct calculation since when $(u_n) \subset H_0^1(0, 1)$ converges to some u_0, it means $(\dot{u}_n)$ converges to $\dot{u}_0$ in $L^2(0, 1)$. Hence, there is some function $g \in L^2(0, 1)$ such that $|\dot{u}_n(t)| \leq g(t)$ for a.e. $t \in [0, 1]$. The application of the Lebesgue Dominated Convergence Theorem proves the assertion. In order to prove the coercivity, we use the following estimation

$$F(u) = \int_0^1 |\dot{u}(t)|^2 dt - 2 \int_0^1 \dot{u}(t) dt + 1 \geq$$
$$\int_0^1 |\dot{u}(t)|^2 dt - 2\sqrt{\int_0^1 |\dot{u}(t)|^2 dt} + 1 \text{ for all } u \in H_0^1(0, 1).$$

Since F is convex and continuous, it is also sequentially weakly lower semi-continuous. Thus, by Theorem 4.3 functional F has exactly one minimizer u_0. Writing down the Euler–Lagrange equation we see that

$$\ddot{u}_0(t) = 0 \text{ and } u(0) = u(1) = 0.$$

Thus $u_0(t) = 0$. Note that $F(u_0) = 1$.

▶ **Remark 4.3** Note that if we consider functional given by (4.6) over $C^1[0, 1]$, we see that the function $u_0(t) = t - 1$ solves the corresponding Euler-Lagrange equation. We prove that this is the global minimizer over $C^1[0, 1]$ by using the fact that $F(u_0) = 0$ and that $F(u) \geq 0$ for all $u \in C^1[0, 1]$. This approach is similar in how one finds extrema in calculus courses: one equates the derivative to zero and next applies some sufficient condition. Note that in the space $C^1[0, 1]$ we cannot use Theorem 4.3 due to the fact that this is not reflexive.

Exercise 4.14

Examine the existence of minimizers from Example 3.3 over $L^2(0, 1)$.

Example 4.8

Consider functional $F : H_0^1(0, 1) \to \mathbb{R}$ defined by the formula

$$F(u) = \int_0^1 \dot{u}^2(t) - 4\pi^2 u^2(t) \, dt. \tag{4.7}$$

We will show that for functional F the Euler-Lagrange equation is solvable by the unique u_0 which is not a minimizer. It is immediate to show that F is continuously differentiable and sequentially weakly lower semicontinuous. The Euler–Lagrange equation reads

$$-\ddot{u}(t) = 4\pi^2 u(t)$$

and is supplied with the boundary conditions

$$u(0) = u(1) = 0.$$

Hence its solution, which we calculate as it is common for second order ODE with constant parameters, is

$$u_0(t) = \sin 2\pi x$$

and $F(u_0) = 0$.

Functional F is not coercive since taking a sequence $u_n(t) = n \sin(2\pi t)$ of functions from $H_0^1(0, 1)$ we easily calculate that $F(u_n) = 0$ and $\|u_n\| \to \infty$ as $n \to +\infty$. Let us take a sequence of functions $v_n \in H_0^1(0, 1)$ defined by

$$v_n(t) = \begin{cases} nt, & t \in \left[0, \frac{1}{2}\right], \\ n - nt, & t \in \left[\frac{1}{2}, 1\right]. \end{cases}$$

Then

$$F(v_n) \to -\infty \text{ as } n \to +\infty$$

and therefore F does not have a minimizer.

Finally, we proceed to the example of a functional that is bounded from below but which does not attain a minimizer. The space over which we consider the problem is not reflexive.

Example 4.9

Consider the following functional $F : C^1[-1, 1] \to \mathbb{R}$ defined by the formula

$$F(u) = \int_{-1}^{1} (t\dot{u}(t))^2 \, dt.$$

which is obviously bounded from below. By a direct argument using the norm convergence in $C^1[-1, 1]$ which is the uniform convergence of functions and their derivatives we reach the conclusion that F is continuous. Consider set

$$D = \left\{ u \in C^1[-1, 1] : u(-1) = 0, \ u(1) = 1 \right\}$$

and sequence $(u_n) \subset D$ defined for $t \in [-1, 1]$ by

$$u_n(t) = \frac{1}{2} + \frac{\arctan\left(\frac{x}{n}\right)}{2\arctan\left(\frac{1}{n}\right)}.$$

Then $F(u_n) \to 0$. Since $F(u) \geq 0$ for $x \in D$, it follows that (u_n) is a minimizing sequence. Therefore, if some $u_0 \in D$ is a minimizer, we obtain that

$$\int_{-1}^{1} (t\dot{u}_0(t))^2 \, dt = 0.$$

Since it implies that $(t\dot{u}_0(t))^2 = 0$, we see that $u_0 = const$, which is not possible since constant functions do not belong to D.

Example 4.10

Consider functional $F : C^1[0, 1] \to \mathbb{R}$ defined by the formula

$$F(u) = \int_0^1 \left(\left(1 - \dot{u}^2(t)\right)^2 + u^2(t) \right) dt. \tag{4.8}$$

We see that $F(u) \geq 0$ for all $u \in C^1[0, 1]$. It is immediate to show that the following

$$u_n(t) = \int_0^t \text{sgn}(\sin 2\pi n s)\, ds \text{ for } n = 1, 2, \ldots$$

stands for the minimizing sequence, i.e.

$$\lim_{n \to +\infty} F(u_n) = 0.$$

We see that $u_n \rightrightarrows 0$ but $|\dot{u}_n(t)| = 1$ for all but a finite number of $t \in [0, 1]$. Moreover, $F(0) = 1$ and for any nonzero function u we have that $F(u) > 0$. Therefore, there is no function in $C^1[0, 1]$ for which the infimum is attained. Note that we cannot consider this functional in $H_0^1(0, 1)$ since it is not well posed there.

Exercise 4.15

Apply the second-order test to check if F defined by (4.8) is convex. Prove that F is continuous and check the type of its differentiability.

4.4   Applications to Control Problems

We will use the Weierstrass Theorem, Theorem 4.3, in studying some control problems. We invite the reader to work on the details of these examples.

Example 4.11

Let

$$S = \left\{ u \in L^2(0, 1) : \|u\|_{L^2} = 1 \right\}.$$

Consider the following boundary value problem for a fixed $u \in S$

$$\dot{x}(t) = -u^2(t), \quad \text{for a.e. } t \in (0, 1), \tag{4.9}$$

$$x(0) = 1, x(1) = 0. \tag{4.10}$$

We will look for such functions $u \in S$, which minimize the following integral

$$J(u) = \int_0^1 t^2 u^2(t)\, dt.$$

In other words, we will consider **the problem of minimization of the functional J over S.**
We proceed as follows. We see that (4.9) is solved by

$$x(t) - c = -\int_0^t u^2(t)\, dt, \ t \in [0, 1].$$

Using relation (4.10) we find that $c = 1$. We see that

$$\inf_{u \in S} J(u) \geq 0.$$

We recall that S is not sequentially weakly compact. Let us take sequence $(u_n) \subset S$ defined by

$$u_n(t) = \begin{cases} n, & t \in [0, \frac{1}{n^2}), \\ 0, & t \in \left[\frac{1}{n^2}, 1\right] \end{cases}$$

which is convergent a.e. to 0 on $[0, 1]$. Since

$$\int_0^1 v(t)\, u_n(t) = \frac{1}{n^2} \int_0^{\frac{1}{n^2}} v(t)\, dt \to 0, \ \text{for each } v \in L^2(0, 1)$$

we have that $u_n \rightharpoonup 0 \notin S$. Moreover,

$$J(u_n) = \int_0^1 t^2 n^2 dt = n^2 \frac{1}{3} \left(\frac{1}{n^2}\right)^3 \to 0 \text{ as } n \to +\infty.$$

From the aforementioned, we see that the given problem is not solvable over S.

Exercise 4.16

Check if the problem from Example 4.11 is solvable if we replace the sphere with a closed unit ball.

Example 4.12

Let $n, m \in \mathbb{N}$ and let $A \in \mathbb{R}^{n \times n}$, $B \in \mathbb{R}^{n \times m}$ be matrices. Assume that $u \in L^2(0, 1; \mathbb{R}^m)$ and consider the following system of differential equations

$$\dot{x}(t) = Ax(t) + Bu(t) \tag{4.11}$$

understood a.e. on $[0, 1]$ with an initial value

$$x(1) = x_0.$$

We understand the solution (4.11) in the sense of Carathéodory, i.e., as an absolutely continuous function over $[0, 1]$. Such a system is solvable by the following Cauchy formula

$$x(t) = x_0 + \int_0^t e^{A(t-s)} Bu(s) ds, \ t \in [0, 1]$$

with any fixed $u \in L^2(0, 1)$. Here $e^{A(t-s)}$ stands for the exponential matrix, whose direct calculation is well described in Chapter 2 [42]. We put

$$D = \left\{ u \in L^2\left(0, 1; \mathbb{R}^m\right) : \|u\|_{L^2} \leq 1 \right\}.$$

Consider set S consisting of pairs (u, x_u), where x_u is a solution to (4.11) with initial condition $x(1) = x_0$ corresponding to $u \in D$.
Let $g : \mathbb{R}^n \to \mathbb{R}$ be a continuous convex function and assume that $h : \mathbb{R}^m \to \mathbb{R}$ is convex and Lipschitz with a constant $L > 0$, i.e.

$$|h(u) - h(v)| \leq L|u - v| \text{ for all } u, v \in \mathbb{R}^n.$$

Our problem is to minimize functional

$$J(u, x) = \int_0^1 (g(x(t)) + h(u(t)))$$

over S.
Using the structure of S we see that we have to minimize

$$J(u) = \int_0^1 \left(g \left(x_0 + \int_0^1 e^{A(t-s)} Bu(s) ds \right) + h(u(t)) \right) dt$$

over D (which is sequentially weakly compact). Thus, we need to show that J is continuous and convex in order to apply Theorem 4.2.
Since we have to prove the convexity of the integral functional, it suffices to show that the integrand is convex. Let us choose arbitrary $u_1, u_2 \in S$ and $\lambda \in [0, 1]$. We obtain what follows for any $t \in [0, 1]$:

$g(x_0 + L(\lambda u_1 + (1 - \lambda)u_2)(t)) = g(x_0 + \lambda L(u_1)(t) + (1 - \lambda)L(u_2)(t)) =$
$g(\lambda [x_0 + L(u_1)(t)] + (1 - \lambda)[x_0 + L(u_2)(t)]) \leq$
$\lambda g(x_0 + L(u_1)(t)) + (1 - \lambda)g(x_0 + L(u_2)(t))$

and also

$$h(\lambda u_1(t) + (1 - \lambda)u_2(t)) \leq \lambda h(u_1(t)) + (1 - \lambda)h(u_2(t)).$$

Therefore, J is also convex.

We need to prove that J is continuous. We observe that mapping $L : D \to AC[0,1]$

$$L(u)(t) = \int_0^1 e^{A(t-s)} Bu(s)ds$$

is linear and continuous over $L^2(0,1)$. Indeed,

$$\|L(u)\|_{AC[0,1]} = \left\|\int_0^1 e^{A(\cdot-s)} Bu(s)ds\right\|_{AC[0,1]} \leq c_1 \|u\|_{L^2(0,1)}$$

for some positive constant c_1.

Let us take sequence $(u_n)_{n\in\mathbb{N}} \subset S$ converging in $L^2(0,1)$ to some $\bar{u} \in D$. Then we obtain

$$|J(u_n) - J(\bar{u})| \leq \int_0^1 |g(x_0 + L(u_n)(t)) - g(x_0 + L(\bar{u})(t))| \, dt + \int_0^1 |h(u_n(t)) - h(\bar{u}(t))| \, dt. \tag{4.12}$$

Since h is Lipschitz by the Schwarz Inequality it follows

$$\int_0^1 |h(u_n(t)) - h(\bar{u}(t))| \, dt \leq L \int_0^1 |(u_n(t)) - (\bar{u}(t))| \, dt \leq L \|u_n - \bar{u}\|_{L^2(0,1)}.$$

This implies that the second term in (4.12) converges to 0. Concerning the first term we see that for a.e. $t \in [0,1]$

$$\lim_{n\to+\infty} g(x_0 + L(u_n)(t)) = g(x_0 + L(\bar{u})(t)).$$

Since L transforms function from D into $AC[0,1]$ and since it is bounded, it follows that the first term in (4.12) converges to 0 due to the Lebesgue Dominated Convergence Theorem.

Therefore, Theorem 4.2 says that the problem under consideration is solvable.

▶ **Remark 4.4** The literature on control problems is very abundant. An interested reader may wish to consult [4, 30] for a detailed mathematical background and also [35, 44]. Some approaches from the area of control theory are to be found in [32] and [10].

4.5   Applications to Second-Order Dirichlet Problems

We start with defining the formal second-order differential operator, which is called
:

$$-\frac{d^2}{dt^2} : H_0^1(0, 1) \rightarrow H^{-1}(0, 1)$$

given by

$$\left\langle -\frac{d^2}{dt^2}u, v \right\rangle = \int_0^1 \dot{u}(t)\,\dot{v}(t)\,dt \text{ for } u, v \in H_0^1(0, 1). \tag{4.13}$$

Sometimes we say that formula (4.13) defines *the Laplacian in the weak sense.*

▶ **Remark 4.5** Note that the negative Laplacian sends points from $H_0^1(0, 1)$ into
linear and continuous functionals defined on $H_0^1(0, 1)$, i.e., elements from $H^{-1}(0, 1)$.
The linearity is obvious. The continuity follows now from the boundedness, which
is a result of the application of the Schwarz Inequality.

Let us impose the following assumption:

A1 $g : [0, 1] \times \mathbb{R} \rightarrow \mathbb{R}$ *is an* L^1-*Carathéodory function.*

We define $G : [0, 1] \times \mathbb{R} \rightarrow \mathbb{R}$ by

$$G(t, x) = \int_0^x g(t, s)\,ds \text{ for a.e. } t \in [0, 1] \text{ and all } x \in \mathbb{R} \tag{4.14}$$

and we see that $\frac{d}{dx}G(t, x) = g(t, x)$ for a.e. $t \in [0, 1]$ and all $x \in \mathbb{R}$.

Exercise 4.17

Prove that G defined by (4.14) is an L^1-Carathéodory function as well.

Under assumption **A1**, we consider the Dirichlet Problem: find a function $u \in$
$H_0^1(0, 1)$ such that the following equation is satisfied:

$$\begin{cases} -\ddot{u}(t) + g(t, u(t)) = 0, \text{ for a.e. } t \in (0, 1), \\ \qquad\qquad u(0) = u(1) = 0. \end{cases} \tag{4.15}$$

Writing that we consider (4.15) in the space $H_0^1(0, 1)$ means that we look for a **weak solution**, i.e., such $u \in H_0^1(0, 1)$ that the following equality holds

$$\int_0^1 \dot{u}(t)\dot{h}(t)dt + \int_0^1 g(t, u(t))h(t)dt = 0 \qquad (4.16)$$

for any $h \in H_0^1(0, 1)$. Note that the definition of the weak solution arises from the definition of the negative Laplacian. We obtain 4.16) formally by assuming that the equality in 4.15) makes sense, i.e., that $u \in H_0^1(0, 1)$ and u has an integrable second order a.e. derivative and next by multiplying both sides of (4.15) by a test function, then taking integrals and integrating by parts. Our approach relies on finding a weak solutions and next on investigating, via the du Bois–Reymond Lemma, its further regularity. By a **classical solution** to (4.15) we understand a function $u \in H_0^1(0, 1)$ which has a second order derivative for a.e. $t \in [0, 1]$ such that $\ddot{u} \in L^1(0, 1)$.

Exercise 4.18

Assume that condition **A1** is satisfied and assume that $u \in H_0^1(0, 1)$ is a weak solution to (4.15). Prove that it is a classical solution to (4.15).

Exercise 4.19

Consider the following Dirichlet problem

$$-\ddot{u}(t) = 0, \quad u(0) = g(0), \quad u(1) = g(1)$$

where $g : \{0, 1\} \to \mathbb{R}$ is a given function. Prove that the function

$$u_0(t) = g(0) + g(1)t$$

is a solution. Show that for $u \in \{v \in H^1(0, 1) : v(0) = g(0), \quad v(1) = g(1)\}$ the following assertions are equivalent:

(a) $u = u_0$.
(b) u is the unique element that has minimal energy "norm" $\|\dot{u}\|_{L^2}$.

Exercise 4.20

Let $f \in L^2(0, 1)$ be fixed. Show that the solution u_0 of the Poisson problem

$$-\ddot{u}(t) = f(t), \quad u(0) = u(1) = 0$$

is the unique minimizer in $H_0^1(0, 1)$ of the quadratic functional

$$J(u) = \frac{1}{2} \int_0^1 |\dot{u}(t)|^2 \, dt - \int_0^1 u(t) f(t) \, dt.$$

Show that $u_0 \in H_0^1(0, 1) \cap H^2(0, 1)$. Prove that J is continuously differentiable.

Additionally we will assume that

A2 *there exist functions* $a \in L^\infty(0, 1)$, b, $c \in L^1(0, 1)$ *such that*

$$\|a\|_{L^\infty} < \pi^2$$

and that

$$G(t, x) \geq -\frac{1}{2} a(t) x^2 + b(t) x + c(t)$$

for a.e. $t \in [0, 1]$ *and all* $x \in \mathbb{R}$.

▶ **Remark 4.6** Assumption $\|a\|_{L^\infty} < \pi^2$ is connected with *the Poincaré Inequality* and is required in order to prove the coercivity of the corresponding Euler action functional $J : H_0^1(0, 1) \to \mathbb{R}$ given by

$$J(u) = \frac{1}{2} \int_0^1 |\dot{u}(t)|^2 \, dt + \int_0^1 G(t, u(t)) \, dt. \tag{4.17}$$

The action functional J is as usually being derived basing on the formula for the weak solution.

Exercise 4.21

Show that functional J given by (4.17) is well defined.

Exercise 4.22

Show that functions

$$G(t, x) = -\frac{1}{4} \pi^2 t x^2 + (\sin t) x, \ G(t, x) = -x^2 \exp(x) + f(t) x$$

serve as examples of a function satisfying **A2,** where $f : \mathbb{R} \to [0, +\infty)$ is some L^1 function.

In order to apply the Direct Method, Theorem 4.3, once we *determine the form of the action functional from the formula defining the weak solution,* we need to proceed according to the following scheme and check:

- *the sequential weak lower semicontinuity of the action functional;*
- *the Gâteaux differentiability of the action functional;*
- *the coercivity of the action functional;*
- *the strict convexity of the action functional (if one wishes to obtain uniqueness).*

In the following sequence of lemmas, we show that the aforementioned are satisfied.

Lemma 4.2

*Assume that condition **A1** holds. Then functional J is sequentially weakly lower semicontinuous on* $H_0^1(0, 1)$.

Proof Observe that $J = J_1 + J_2$, where

$$J_1(u) = \frac{1}{2} \int_0^1 |\dot{u}(t)|^2 \, dt \text{ and } J_2(u) = \int_0^1 G(t, u(t)) \, dt.$$

Functional J_1 is convex and continuous, which by Theorem 2.3 implies that it is sequentially weakly lower semicontinuous. From Example 2.8 we see that J_2 is sequentially weakly continuous, which further implies that J is sequentially weakly lower semicontinuous.

Lemma 4.3

*Assume that condition **A1** holds. Then functional J is continuously differentiable on* $H_0^1(0, 1)$ *and for any fixed* $u \in H_0^1(0, 1)$ *the derivative reads:*

$$\left\langle J'(u), h \right\rangle = \int_0^1 \dot{u}(t)\dot{h}(t) + g(t, u(t))h(t) dt \text{ for all } h \in H_0^1(0, 1).$$

Moreover, critical points to J are weak solutions to 4.15).

Proof From Example 3.12 it follows that J_1 is a C^1 functional. Example 3.18 tells us that J_2 is also C^1. It is immediate to see that critical points to J are weak solutions to 4.15).

Lemma 4.4

*Under assumptions **A1**, **A2** functional J is coercive over $H_0^1(0, 1)$.*

Proof Put

$$a_1 = \|a\|_{L^\infty}, \ b_1 = \int_0^1 |b(t)| \, dt, \ c_1 = \int_0^1 c(t) dt.$$

Observe that by the Sobolev and the Poincaré Inequality:

$$J_2(u) \geq -\tfrac{1}{2} \int_0^1 a(t) u^2(t) \, dt + \int_0^1 b(t) u(t) \, dt + \int_0^1 c(t) dt \geq$$
$$-\tfrac{1}{2\pi^2} a_1 \|u\|_{H_0^1}^2 - b_1 \|u\|_{H_0^1} + c_1 \text{ for any } u \in H_0^1(0, 1).$$

Summing up we have

$$J(u) \geq \frac{1}{2} \left(1 - \frac{1}{\pi^2} a_1\right) \|u\|_{H_0^1}^2 - b_1 \|u\|_{H_0^1} - c_1 \text{ for any } u \in H_0^1(0, 1).$$

Since $a_1 < \pi^2$, we see that J is coercive over $H_0^1(0, 1)$.

Using Lemmas 4.2, 4.3 and 4.4 we can apply Theorem 4.3 to reach the following result:

Theorem 4.7

*Assume conditions **A1**, **A2**. Then problem (4.15) has at least one classical solution.*

Exercise 4.23

Prove that if assume **A1** and if we replace **A2** with the following: *there exist* $a \in L^\infty(0, 1)$, $b, c \in L^1(0, 1)$, *a number* $\beta \in (1, 2)$ *such that*

$$G(t, x) \geq -\frac{1}{2} a(t) |x|^\beta + b(t) x + c(t), \text{ for a.e. } t \in [0, 1] \text{ and all } x \in \mathbb{R}$$

then (4.15) has at least one classical solution $u \in H_0^1(0, 1)$. Hint: Show that under the new assumption functional J is coercive for any $a \in L^\infty(0, 1)$. Can the assumptions on function a be relaxed?

It remains to comment that the uniqueness is reached in case functional J has exactly one critical point. Hence, we need one additional assumption:

A3 *for a.e. $t \in [0, 1]$ function $x \mapsto g(t, x)$ is non-decreasing on $\mathbb{R}$.*

Theorem 4.8

*Assume that conditions **A1**, **A2**, **A3** hold. Then Dirichlet Problem (4.15) has exactly one classical solution $u \in H_0^1(0, 1)$.*

Proof Using Example 3.8 and exercise that follows we see that J_2 is convex. Since J_1 is strictly convex, we see that now J is strictly convex and therefore its critical point, which exists by Theorem 4.7, is unique.

Exercise 4.24

Assume that conditions **A1**, **A2**, **A3** hold and that additionally function g is (jointly) continuous. Show that 4.15 has exactly one solution $u \in H_0^1(0, 1) \cap C^2[0, 1]$.

Now, via a number of exercises, we investigate some special cases of problem (4.15). In order to solve them, the reader may use the pattern just described earlier together with some earlier introduced methods and results.

Exercise 4.25

Let $g : \mathbb{R} \to \mathbb{R}$ be a bounded continuous function. Consider the Dirichlet problem

$$\begin{cases} -\ddot{u}(t) + g(u(t)) = 0, & \text{for } t \in (0, 1), \\ u(0) = u(1) = 0. \end{cases}$$

Prove that the aforementioned problem has at least one solution $u \in H_0^1(0, 1) \cap C^2[0, 1]$.

Exercise 4.26

Let $g : \mathbb{R} \to \mathbb{R}$ be a continuous and nondecreasing function. Consider the Dirichlet problem

$$\begin{cases} -\ddot{u}(t) + g(u(t)) = 0, & \text{for } t \in (0, 1), \\ u(0) = u(1) = 0. \end{cases}$$

Prove that the aforementioned problem has exactly one solution $u \in H_0^1(0, 1) \cap C^2[0, 1]$.

Exercise 4.27

Let $J : H_0^1(0, 1) \to \mathbb{R}$ be defined by

$$J(u) = \frac{1}{2} \int_0^1 |\dot{u}(t)|^2 \, dt - \frac{\alpha}{2} \int_0^1 |u(t)|^2 \, dt.$$

Prove that for all $\alpha < \pi^2$ functional J is strictly convex.

From the aforementioned exercise, it follows that we can weaken the assumptions of Theorem 4.8 as far as the uniqueness is concerned. We provide thereby a type of relaxed convexity, which is of some importance.

Exercise 4.28

Assume that conditions **A1**, **A2** hold. Assume also that for some $0 < \alpha < \pi^2$ function

$$x \mapsto \frac{\alpha}{2} x^2 + G(t, x)$$

is convex on $\mathbb{R}$ for a.e. $t \in [0, 1]$. Then Dirichlet Problem 4.15) has exactly one classical solution $u \in H_0^1(0, 1)$. Hint: Write the action functional as follows

$$J(u) = \left(\frac{1}{2} \int_0^1 |\dot{u}(t)|^2 \, dt - \frac{\alpha}{2} \int_0^1 |u(t)|^2 \, dt \right)$$
$$+ \left(\frac{\alpha}{2} \int_0^1 |u(t)|^2 \, dt + \int_0^1 G(t, u(t)) \, dt \right).$$

4.6   On the Best Constant in the Poincaré Inequality

Now we proceed to the derivation of the precise value of the constant in the general Poincaré Inequality, i.e.

$$\|u\|_{L^2} \leq c \|\dot{u}\|_{L^2} \quad \text{for } u \in H_0^1(0, 1) \tag{4.18}$$

for some constant $c \in (0, 1]$. The upper bound on c, which equals 1, is taken after Remark 2.6. Now we proceed as in [28], Section 14.5*, with necessary modifications due to the way in which we understand the negative Laplacian and due to the exposition of this text. We start with the following exercise, which provides a direct formula for the solution of the Poisson problem:

Exercise 4.29

Let $k : [0, 1] \times [0, 1] \to \mathbb{R}$ be defined by

$$k(t, s) = \begin{cases} (1 - t)\,s, & 0 \le s \le t, \\ t\,(1 - s), & t < s \le 1. \end{cases}$$

Prove that:

(a) the function k is continuous;
(b) $\|k\|_\infty \le 1$;
(c) $k(t, s) = k(s, t)$ for all $s, t \in [0, 1]$;
(d) for any function $f \in C[0, 1]$ function $h : [0, 1] \to \mathbb{R}$ defined by

$$h(t) = \int_0^1 k(t, s)\, f(s)\, ds \qquad (4.19)$$

is twice continuously differentiable, $h(0) = h(1) = 0$ and $\ddot{h}(t) = -f(t)$ for $t \in [0, 1]$. Hint: Observe that (4.19) can be equivalently written as

$$h(t) = -\int_0^t (t - s)\, f(s)\, ds + t \int_0^1 (1 - s)\, f(s)\, ds. \qquad (4.20)$$

and apply *the Leibniz Rule*.
(e) examine the situation whether (iv) holds with the assumption that $f \in L^2(0, 1)$. How the differentiability is understood in this case?

Exercise 4.30

Show that the space $H^2(0, 1) \cap H_0^1(0, 1)$ is dense in $H_0^1(0, 1)$. Next show that *the Poincaré Inequality* is equivalent to

$$\|u\|_{L^2} \le c \|\dot{u}\|_{L^2} \quad \text{for } u \in H^2(0, 1) \cap H_0^1(0, 1).$$

The aforementioned exercise and also results from Sect. 4.5 allow us to formulate the following:

Lemma 4.5
The negative Laplacian

$$-\frac{d^2}{dt^2} : H_0^1(0, 1) \to H^{-1}(0, 1)$$

(Continued)

Lemma 4.5 (continued)

given by the formula (4.13) is invertible with an inverse given by the following formula in case $f \in L^2(0, 1)$:

$$\left(-\frac{d^2}{dt^2}\right)^{-1} f = \int_0^1 k(t, s) f(s) \, ds \qquad (4.21)$$

We will denote $\left(-\frac{d^2}{dt^2}\right)^{-1} = A$ for simplicity having in mind that A : $L^2(0, 1) \to H_0^1(0, 1)$.

▶ **Remark 4.7** Due to the continuous and dense inclusion $L^2(0, 1) \subset H^{-1}(0, 1)$ it makes sense to provide inversion formula on $L^2(0, 1)$.

Exercise 4.31

Prove that operator A defined by (4.21) is linear and bounded. Estimate its norm.

Exercise 4.32

Prove that operator A defined by (4.21) is self-adjoint and compact.

Exercise 4.33

Demonstrate that the largest absolute value of an eigenvalue of A reads $1/\pi^2$ and corresponds to the eigenfunction

$$e_1(t) = \frac{\sqrt{2}}{2} \sin(\pi t), \ t \in [0, 1].$$

Use the Spectral Theorem to prove that $\|A\| = 1/\pi^2$ ($\|A\|$ denotes the norm of A). Show also that

$$\left(\|Af\|_{H_0^1}\right)^2 \le \|A\|(Af, f) \quad \left(f \in L^2(a, b)\right). \qquad (4.22)$$

Note that we can rewrite the Poincaré Inequality (4.18) as follows

$$\|u\|_{L^2} \le c\,(u, -\ddot{u}) \quad \text{for } u \in H_0^1(0, 1)$$

given that we can consider it for $u \in H^2(0, 1) \cap H_0^1(0, 1)$. Using the assertion of Lemma 4.5, then we can rewrite the aforementioned inequality further as follows

$$\|Af\|_{L^2} \le c\,(Af, f) \text{ for } f \in L^2(0, 1).$$

Taking the Cauchy–Schwarz Inequality into account, we have

$$(Af, f) < \|Af\| \|f\| \leq \|A\| \|f\|^2$$

which implies that the optimal constant in the Poincaré Inequality satisfies $c_0 \geq \sqrt{\|A\|}$. Using (4.22) we see that

$$c_0 = \sqrt{\|A\|}.$$

4.7   On Some Abstract Formulation of the Direct Method

In this section, we assume that E is a real Hilbert space. We will derive some abstract version of *the Lax–Milgram Lemma* supplied further with the applications meant for the type of problems, which we have considered so far.

Definition 4.2 (Bounded bilinear form)
Bounded bilinear form on E is a functional

$$a : E \times E \rightarrow \mathbb{R}$$

fulfilling the following conditions:

(a) for all $x, y, z \in E, \alpha, \beta \in \mathbb{R}$

$$a(\alpha x + \beta y, z) = \alpha a(x, z) + \beta a(y, z),$$

$$a(z, \alpha x + \beta y) = \alpha a(z, x) + \beta a(z, y);$$

(b) there exists a constant $d > 0$ such that

$$|a(x, y)| \leq d \|x\| \|y\| \text{ for } x, y \in E.$$

Definition 4.3
Bounded bilinear form $a : E \times E \rightarrow \mathbb{R}$ is called

(a) *symmetric* if and only if

$$a(x, y) = a(y, x) \text{ for } x, y \in E.$$

(b) *positive* if and only if

$$a(x, x) \geq 0 \text{ for } x \in E.$$

for all $u \in E$; a is called *strictly positive* if and only if

$$a(x, x) > 0 \text{ for } x \in E, \ x \neq 0.$$

(c) *strongly positive* if and only if there exists a constant $c > 0$ such that

$$a(x, x) \geq c \|x\|^2 \text{ for } x \in E.$$

Exercise 4.34

Prove that the strongly positive form is strictly positive.

Exercise 4.35

Prove that a bounded bilinear form $a : E \times E \to \mathbb{R}$ is continuos.

Exercise 4.36

Assume that $a : E \times E \to \mathbb{R}$ is a bounded and symmetric bilinear form. Prove that the functional $J : E \to \mathbb{R}$ defined by

$$J(x) = \frac{1}{2} a(x, x)$$

is continuously differentiable and that J has the second order Gateaux variation given for any $x \in E$ by the formula

$$J^{(2)}(x; h) = a(h, h) \text{ for } h \in E.$$

Example 4.13

It is easy to check that in any real Hilbert space there is a bounded symmetric strongly positive bilinear form. In particular in $H_0^1(0, 1)$ we have

$$a(u, v) = \int_0^1 \dot{u}(t) \, \dot{v}(t) \, dt \text{ for } u, v \in H_0^1(0, 1)$$

and we see that $c = 1$ and $d = 1$.

Now we proceed with some version of *the Lax–Milgram Theorem* (see also Corollary 5.8 from [8]), which finds many applications in the solvability of boundary value problems for differential equations (partial including):

Theorem 4.9 (Lax–Millgram)

Assume that $a : E \times E \to \mathbb{R}$ is a bounded and strongly symmetric positive bilinear form. Then for any given element $b \in E^*$ there exists a unique x_0 such that

$$a(x_0, h) = \langle b, h \rangle \text{ for all } h \in E. \tag{4.23}$$

Moreover, x_0 is the unique minimizer to the following functional

$$J(x) = \frac{1}{2} a(x, x) - \langle b, x \rangle, \quad J : E \to \mathbb{R}.$$

Proof It is easy to check that J is C^1. Functional J is coercive since

$$J(x) \geq \frac{1}{2} c \|x\|^2 - \|b\|_* \|x\| \text{ for all } x \in E.$$

As for the second variation, we see that

$$J^{(2)}(x; h) = a(h, h) \geq \frac{1}{2} c \|h\|^2 > 0$$

for all $h \neq 0$, $h \in E$ and all $x \in E$. Summarizing, functional J is strictly convex, sequentially weakly lower semicontinuous and coercive. Therefore, J has exactly one minimizer u_0, which satisfies (4.23).

Example 4.14 (Application to Dirichlet Problems with Fixed Right-Hand Side)

Put $E = H_0^1(0, 1)$ and consider the following linear Dirichlet problem

$$\begin{cases} -\ddot{u}(t) + h(t) = 0, & \text{for a.e. } t \in (0, 1), \\ u(0) = u(1) = 0, \end{cases} \tag{4.24}$$

where $h \in L^2(0, 1)$ is some fixed function. Observe that h defines a continuous linear functional $b : E \to \mathbb{R}$ given by

$$b(v) = \int_0^1 v(t) h(t) \, dt \text{ for } v \in E.$$

We already know from Example 4.13 that $a : H_0^1(0, 1) \times H_0^1(0, 1) \to \mathbb{R}$ given by

$$a(u, v) = \int_0^1 \dot{u}(t) \dot{v}(t) \, dt$$

is a bounded and strongly symmetric positive bilinear form. Thus, Theorem 4.9 says that there is exactly one $u_0 \in H_0^1(0, 1)$ which is a weak solution to (4.24), that is

$$\int_0^1 \dot{u}_0(t)\,\dot{v}(t)\,dt = \int_0^1 v(t)\,h(t)\,dt \text{ for all } v \in H_0^1(0, 1).$$

Moreover the application of the du Bois–Reymond Lemma implies that $u_0 \in H^2(0, 1) \cap H_0^1(0, 1)$.

Exercise 4.37

Can we obtain the results from Example 4.14 with $h \in L^1(0, 1)$?

In order to consider nonlinear Dirichlet problems of the type 4.15), we provide another abstract result:

Theorem 4.10

Assume that $a : E \times E \to \mathbb{R}$ is a bounded and strongly symmetric positive bilinear form and that $b : E \to \mathbb{R}$ is a concave C^1 functional such that there are constants $\alpha < c$, b, $\gamma \in \mathbb{R}$ for which

$$b(x) \le \frac{1}{2}\alpha \|x\|^2 + \beta \|x\| + \gamma \text{ for all } x \in E. \tag{4.25}$$

Then the Euler action functional $J : E \to \mathbb{R}$ defined by

$$J(x) = \frac{1}{2}a(x, x) - b(x)$$

is C^1 with the following derivative for any $x \in E$

$$\left\langle J'(x); h \right\rangle = a(x, h) - \left\langle b'(x), h \right\rangle \text{ for all } h \in E.$$

Moreover, J has exactly one minimizer x_0 such that

$$a(x_0, h) = \left\langle b'(x_0), h \right\rangle \text{ for } h \in E.$$

Proof Since b is continuous and concave, it follows that $-b$ as a convex continuous functional it is also sequentially weakly lower semicontinuous. Therefore, J is sequentially weakly lower semicontinuous and strictly convex. Using (4.25) and relation that $\frac{1}{2}a(x, x) \ge \frac{1}{2}c\|x\|^2$ for all $x \in E$ we get that functional J is additionally coercive since

$$J(x) \geq \frac{1}{2}(c - \alpha) \|x\|^2 - \beta \|x\| - \gamma \text{ for all } x \in E.$$

Thus, we reach the conclusion by the application of Theorem 4.3.

Example 4.15 (Applications to Problem with a Nonlinear Term)

Let $E = H_0^1(0, 1)$. Assume conditions **A1–A3** with the only change that we assume $g : [0, 1] \times \mathbb{R} \to \mathbb{R}$ to be an L^2–Carathéodory function. We consider the nonlinear Dirichlet problem

$$\begin{cases} -\ddot{u}(t) + g(t, u(t)) = 0, & \text{for a.e. } t \in (0, 1), \\ u(0) = u(1) = 0. \end{cases} \tag{4.26}$$

We put

$$b(u) = \int_0^1 G(t, u(t)) \, dt$$

and we see immediately that Theorem 4.8 applies in order to determine the unique solvability of (4.26).

Exercise 4.38

Obtain the results from Example 4.14 with assuming that h is an L^1–Carathéodory function.

Applications to Multiple Integrals

Up to now, we have considered problems related to the field of ODE. Now we show how methods that we have indicated earlier apply in much more advanced problems roughly speaking related to the field of PDE. We need to fix some background about spaces, which we will consider and next proceed to counterparts of examples and of results we have given so far. There is a number of excellent sources treating the case of PDE. We have used the mainly [7] for the function space setting, [43] for some examples and [18] for many comments and explanations.

5.1   Instead of an Introduction

Let us provide some alternative, yet equivalent, definition of the space $H^1(0, 1)$ as given in [8]. We try to be rather short in our presentation directing an interested reader for example to [27] for a more detailed description. We now resort to the notion of the weak derivative, which will further allow us to go smoothly into a more advanced setting. From the theory of L^p spaces we know that $C^1[0, 1]$ is dense in $L^p(0, 1)$ for any $1 \leq p < +\infty$.

Definition 5.1 (Weak Derivative)
We say that a function $u \in L^2(0, 1)$ belongs to $H^1(0, 1)$ if there is a function $v \in L^2(0, 1)$ such that

$$\int_0^1 u(t)\, \dot{\varphi}(t)\, dt = -\int_0^1 v(t)\, \varphi(t)\, dt \qquad (5.1)$$

for all (test functions) $\varphi \in C_0^1[0, 1]$. Function v is called a *weak derivative* of function u.

M. Galewski, *Basics of Nonlinear Optimization*, Compact Textbooks in Mathematics, https://doi.org/10.1007/978-3-031-77160-6_5

Exercise 5.1

Show that a weak derivative is uniquely defined.

If u is absolutely continuous, then (5.1) becomes a formula of integration by parts (see Proposition 2.1) and so v is a classical derivative of u. This is why we denote $\dot{u} = v$ and call it a weak derivative or a derivative in a sense of the space $H^1(0, 1)$. Classical derivatives are defined pointwise, as limits of difference quotients. On the other hand, weak derivatives are defined only in an integral sense, up to a set of measure zero. By arbitrarily changing the function on a set of measure zero, we do not affect its weak derivative.

Example 5.1

A function

$$u(t) = \left| t - \frac{1}{2} \right|, \tag{5.2}$$

which is not differentiable in the classical calculus sense, belongs to $H^1(0, 1)$. We have

$$\dot{u}(t) = \begin{cases} -1, & 0 \leq t < \frac{1}{2}, \\ 1, & \frac{1}{2} < t \leq 1. \end{cases} \tag{5.3}$$

More generally, a continuous function that is piecewise C^1 on $[0, 1]$ belongs to $H^1(0, 1)$.

Example 5.2

Note that function u given by (5.3) does not belong to the space $H^1(0, 1)$, but it has a classical derivative equal to 0 for a.e. $t \in [0, 1]$. The only candidate for a derivative in a sense of space $H^1(0, 1)$ is as expected function $v \equiv 0$ for which formula (5.1) does not hold for all test functions.

Exercise 5.2

Verify that the a derivative of function (5.2) is given by (5.3). Show in detail that function given by 5.3) does not belong to $H^1(0, 1)$.

Example 5.3

We consider the everywhere discontinuous (and thus nowhere differentiable) function $u : [0, 1] \to \mathbb{R}$

$$u(t) = \begin{cases} 0, & \text{if } t \text{ is rational} \\ 2 + \sin t, & \text{if } t \text{ is irrational}. \end{cases}$$

We easily calculate that the function $g(x) = \cos x$ stands for a weak derivative for u. Indeed, the behavior of u on the set of rational points (having measure zero) is irrelevant as already mentioned. We thus have for any test function $\varphi \in C_0^1 [0, 1]$:

$$- \int_0^1 u(t)\dot{\varphi}(t)\mathrm{d}t = - \int_0^1 (2 + \sin x)\dot{\varphi}(t)\mathrm{d}t = \int (\cos x)\varphi(t)\,\mathrm{d}t.$$

Exercise 5.3

Verify whether the sign function

$$\mathrm{sgn}\,(x) = \begin{cases} -1 & \text{if } x < 0 \\ 0 & \text{if } x = 0 \\ +1 & \text{if } x > 0 \end{cases}$$

has a weak derivative.

Example 5.4

Consider the Cantor function $u : \mathbb{R} \mapsto [0, 1]$ (see [15] for a detailed description of such function) defined by

$$u(t) = \begin{cases} 0 & \text{if } t \le 0, \\ 1 & \text{if } t \ge 1, \\ 1/2 & \text{if } t \in [1/3, 2/3], \\ 1/4 & \text{if } t \in [1/9, 2/9], \\ 3/4 & \text{if } t \in [7/9, 8/9], \\ \quad \cdots \end{cases}$$

which provides a standard example of a continuous function, which is not absolutely continuous, as mentioned in Sect. 2.5. Restrict u to $[0, 1]$. We show that such restricted u does not have a weak derivative. Suppose that v stands for the weak derivative. Since u is constant on each of the open sets

$$\left(\frac{1}{3}, \frac{2}{3}\right), \quad \left(\frac{1}{9}, \frac{2}{9}\right), \quad \left(\frac{7}{9}, \frac{8}{9}\right), \cdots$$

we must have $v(t) = \dot{u}(t) = 0$ on the union of these open intervals. Hence $v(t) = 0$ for a.e. $t \in [0, 1]$. It remains to show that v is not the weak derivative of u. Let $\varphi \in C_0^1 [0, 1]$ be a test function such that $\varphi(t) = 1$ for $x \in \left(\frac{1}{3}, \frac{2}{3}\right)$ while $\varphi(t) = 0$ otherwise. Such a test function necessarily exists. Indeed, recalling formula (2.13), we define

$$\varphi(t) = \begin{cases} \left(t - \frac{1}{3}\right)^2 \left(\frac{2}{3} - t\right)^2, & t \in \left[\frac{1}{3}, \frac{2}{3}\right], \\ 0, & t \notin \left[\frac{1}{3}, \frac{2}{3}\right], \end{cases}$$

and thus, we reach a contradiction.

Now we turn to a result that says that elements of $H^1(0, 1)$ are continuous (precisely speaking are equivalent to continuous functions):

Exercise 5.4

Let $u \in L^1(0, 1)$ be such that

$$\int_0^1 u(t)\dot{\varphi}(t)\,dt = 0 \text{ for any } \varphi \in C_0^1[0, 1]. \tag{5.4}$$

Using the definition of the weak derivative, prove that there is a constant $c \in \mathbb{R}$ such that $u(t) = c$ for a.e. $t \in [0, 1]$.

Theorem 5.1

Let $u \in H^1(0, 1)$. Then there is a function $u_0 \in C[0, 1]$ such that

$$u(t) = u_0(t) \text{ for a.e. } t \in [0, 1]$$

and moreover

$$u_0(t) - u_0(s) = \int_s^t \dot{u}(\tau)\,d\tau$$

for any $s, t \in [0, 1]$.

Proof We fix $s \in [0, 1]$ and define function $v : [0, 1] \to \mathbb{R}$ by the following formula

$$v(t) = \int_s^t \dot{u}(\tau)\,d\tau.$$

Then from Lemma 2.4, it follows that

$$\int_0^1 v(t)\dot{\varphi}(t)\,dt = -\int_0^1 \dot{u}(t)\varphi(t)\,dt \text{ for any } \varphi \in C_0^1[0, 1].$$

Hence,

$$\int_0^1 (v(t) - u(t)) \, \dot\varphi(t) \, dt = 0 \text{ for any } \varphi \in C_0^1[0, 1]$$

and the assertion follows if we use the preceding exercise.

▶ **Remark 5.2** The above theorem asserts that every function $u \in H^1(0, 1)$ admits one (and only one) absolutely continuous representative on $[0, 1]$ (see Remark 8, Chapter 8 in [8] for details). Thus we can replace u by its absolutely continuous representative. Thereby we see that we have arrived at equivalent formulation of the Sobolev space $H^1(0, 1)$. Note that we can now shift the approach just presented into the case of space $H^1(\Omega)$, where Ω is some subset of $\mathbb{R}^N$.

5.2 On the Space $C(\Omega)$

Let (Ω, d) be a compact metric space. We can take for example a closed and bounded subset of $\mathbb{R}^N$ equipped with Euclidean distance. By $C(\Omega)$ we denote the space of all continuous real-valued functions $u : \Omega \to \mathbb{R}$. The natural norm is given as follows:

$$\|u\|_\infty = \sup_{x \in \Omega} |u(x)|$$

and the convergence implied by this norm is again called uniform convergence. Since Ω is a compact set, we see due to the Weierstrass Theorem, see Sect. 4, that the supremum is attained and we can replace sup with max .

Lemma 5.1
$C(\Omega)$ *is a Banach space.*

Proof Let (u_n) be a Cauchy sequence in $C(\Omega)$. Then for every fixed $x \in \Omega$ the sequence of numbers $u_n(x)$ is Cauchy and hence converges to some limit, which we call $u(x)$. Thereby we have defined a function $u : \Omega \mapsto \mathbb{R}$ about which we need to prove the boundedness and the continuity. By assumption, for every $\varepsilon > 0$ there exists N_ε so that

$$\|u_n - u_m\|_\infty \le \varepsilon \quad \text{for all } n, m \ge N_\varepsilon \tag{5.5}$$

Letting $m \to \infty$, since $u_m(x) \to u(x)$ for $x \in \Omega$, we obtain

$$\sup_{n \ge N_\varepsilon} \sup_{x \in \Omega} |u_n(x) - u(x)| \le \varepsilon \tag{5.6}$$

which implies

$$\sup_{n \geq N_\varepsilon} \|u_n - u\|_\infty \leq \varepsilon$$

and therefore $\|u_n - u\|_\infty \to 0$. Now we show that u is bounded and continuous. Indeed, from (5.6) and next from (5.5) we learn that

$$\sup_{x \in \Omega} |u(x)| \leq \varepsilon + \sup_{x \in \Omega} |u_n(x) - u_{N_\varepsilon}(x)| + \sup_{x \in \Omega} |u_{N_\varepsilon}(x)| < +\infty$$

which implies that u is bounded.

Next we prove that u is continuous. Let any $x \in \Omega$ and $\varepsilon > 0$ be given. By the uniform convergence, there exists an integer N_ε such that

$$|u_N(x) - u(x)| < \varepsilon/3 \text{ for every } x \in \Omega.$$

Since u_{N_ε} is continuous, there exists $\delta > 0$ such that

$$\left|u_{N_\varepsilon}(y) - u_{N_\varepsilon}(x)\right| < \varepsilon/3 \text{ whenever } d(y, x) < \delta.$$

Putting together the aforementioned inequalities, when $d(y, x) < \delta$ we have

$$|u(y) - u(x)| \leq \left|u(y) - u_{N_\varepsilon}(y)\right| + \left|u_{N_\varepsilon}(y) - u_{N_\varepsilon}(x)\right| + \left|u_{N_\varepsilon}(x) - u(x)\right|$$
$$< \frac{\varepsilon}{3} + \frac{\varepsilon}{3} + \frac{\varepsilon}{3} = \varepsilon,$$

proving that u is continuous at the point x.

A family of continuous functions $\mathcal{F} \subset C(\Omega)$ is called equi-continuous if, for every $x \in \Omega$ and $\varepsilon > 0$, there exists $\delta > 0$ such that

$$d(y, x) < \delta \text{ implies } |u(y) - u(x)| < \varepsilon$$

for all functions $u \in \mathcal{F}$. The idea of equicontinuity says that $\delta > 0$ can depend on x and ε, but not on the particular function $u \in \mathcal{F}$. For a subset $\mathcal{F} \subset C(\Omega)$ the following properties are equivalent:

(i) $\mathcal{F}$ is relatively compact, i.e., the closure $\overline{\mathcal{F}}$ is compact.
(ii) From every sequence of continuous functions $(u_n) \subset \mathcal{F}$, one can extract a subsequence converging to some function u, uniformly on Ω.

Theorem 5.2 (Arzela-Ascoli)
Let $\mathcal{F} \subset C(\Omega)$ be an equi-continuous family of functions, such that

$$\sup_{u \in \mathcal{F}} |u(x)| < \infty \quad \text{for every } x \in \Omega.$$

Then $\mathcal{F}$ is a relatively compact subset of $C(\Omega)$.

Let now $\Omega \subseteq \mathbb{R}^N$ be an open and bounded set. Then $\overline{\Omega}$ is compact and we can consider the space $C\left(\overline{\Omega}\right)$ just as described above. We introduce also the space $C^1\left(\overline{\Omega}\right)$ consisting of real valued functions which are differentiable on some open neighborhood V of $\overline{\Omega}$ and whose first order partial derivatives are continuous on $\overline{\Omega}$. The norm in $C^1\left(\overline{\Omega}\right)$ is defined similarly to its counterpart given for C^1 functions considered on $[0, 1]$.

Exercise 5.5

Prove that formulas

$$\|u\|_1 = \sup_{x \in \overline{\Omega}} \left(|u(x)| + |\nabla u(x)|\right), \; \|u\|_2 = \sup_{x \in \overline{\Omega}} |u(x)| + \sup_{x \in \overline{\Omega}} |\nabla u(x)|$$

$$\text{and } \|u\|_3 = \sup_{x \in \overline{\Omega}} |\nabla u(x)| + |u(x_0)|$$

with some fixed $x_0 \in \overline{\Omega}$ make $C^1\left(\overline{\Omega}\right)$ into a Banach space.

We distinguish between $C\left(\overline{\Omega}\right)$ and $C\left(\Omega\right)$ as well as between $C^1\left(\overline{\Omega}\right)$ and $C^1\left(\Omega\right)$ as mentioned in the case of functions defined on the real line.

5.3   Sobolev Spaces

Let $\Omega \subset \mathbb{R}^N$ be an open unit ball. We would like to underline that results concerning Sobolev spaces deeply rely on the properties of the domain as well as on its boundary. Our aim is to illustrate the optimization and variational tools that we use rather than to discuss the theory of Sobolev spaces in detail. Therefore we will not concentrate on many subtle details. One may consider an open, connected and bounded set with sufficiently regular boundary, see [18] and [43] for details. We direct the Reader to the books [1, 7, 18] and [43] for a through discussion. Let $D \subset \mathbb{R}^N$ be a measurable set. By $L^1_{\text{loc}}(D)$ we denote the space of locally summable functions on D, i.e. these are the measurable functions $u : D \mapsto \mathbb{R}$ which are summable when restricted to every compact subset $K \subset D$.

Example 5.5

The functions e^x and $\ln(|x| + 1)$ are in $L_{loc}^1(\mathbb{R})$, while obviously $x^{-1} \notin L_{loc}^1(\mathbb{R} \setminus \{0\})$. On the other hand, the function $u(x) = x^\gamma$ is in $L_{loc}^1(0, \infty)$ for every (positive or negative) exponent $\gamma \in \mathbb{R}$. In several space dimensions (as well as in the case $N = 1$), the function $u(x) = |x|^{-\gamma}$ is in $L_{loc}^1(\mathbb{R}^N)$ provided that $\gamma < N$. It should be recalled that the pointwise values of a locally summable function on a set of measure zero are irrelevant.

By $\mathcal{C}_c^\infty(\Omega)$ we denote the space of continuous functions $\varphi : \Omega \to \mathbb{R}$ having continuous partial derivatives of all orders and whose support is a compact subset of Ω. Functions $\varphi \in \mathcal{C}_c^\infty(\Omega)$ are usually called "test functions". We recall that the support of a function φ is the closure of the set where φ does not vanish:

$$\text{Supp}(\varphi) = \overline{\{x \in \Omega; \varphi(x) \neq 0\}}.$$

In accordance to the weak derivative, we will introduce weak partial derivatives of a function of several variables and also the notion of the weak gradient.

Definition 5.2 (Weak Derivative)

Let $u \in L_{loc}^1(\Omega)$ be a locally summable function. If for some $1 \leq i \leq N$ there exists a locally summable function $v \in L_{loc}^1(\Omega)$ such that

$$\int_\Omega u(x) \frac{\partial \varphi(x)}{\partial x_i} dx = - \int_\Omega v(x) \varphi(x) dx \quad \text{for all test functions } \varphi \in \mathcal{C}_c^\infty(\Omega)$$

then we say that v is the weak partial derivative of u with respect to variable x_i and write

$$v = \frac{\partial u}{\partial x_i} \text{ or } v = u_{x_i}.$$

If all weak partial derivatives exist, we say that function u is weakly differentiable and define the weak gradient as follows

$$\nabla u = \left[\frac{\partial}{\partial x_1} u, \ldots, \frac{\partial}{\partial x_N} u \right].$$

Exercise 5.6

Show that the weak partial derivative, if it exists, is defined uniquely (up to a set of measure zero).

Exercise 5.7

Show that if $u \in C^1(\Omega)$, then partial derivatives and weak partial derivatives coincide. How about the case when u has merely partial derivatives which need not be continuous?

Example 5.6

If we consider the function $u : \mathbb{R}^2 \to \mathbb{R}$

$$u(x, y) = \begin{cases} 0 & \text{if } x \le 0, \ y \in \mathbb{R}, \\ x & \text{if } x > 0, \ y \in \mathbb{R} \end{cases}$$

then we easily check that the Heaviside type function $H \in L^1(\mathbb{R}^2)$ defined as follows

$$H(x, y) = \begin{cases} 0 & \text{if } x \le 0, \ y \in \mathbb{R} \\ 1 & \text{if } x > 0, \ y \in \mathbb{R} \end{cases}$$

is the weak derivative of u with respect to x, while the weak derivative with respect to y is 0. Note that function H itself does not admit a weak derivative with respect to x since the only candidate is the function equal constantly to 0. Then arguing as in the Example 5.4 we arrive at a contradiction.

Exercise 5.8

Calculate the weak partial derivatives of the following function

$$u(x, y) = |x| \, y + \frac{1}{2} y^2.$$

Exercise 5.9

Show that if $u, v \in L^1_{loc}(\Omega)$ are weakly differentiable functions, then for any constants $a, b \in \mathbb{R}$ the linear combination $af + bv$ is also weakly differentiable and

$$\frac{\partial}{\partial x_i}(af + bv) = a \frac{\partial u}{\partial x_i} + b \frac{\partial v}{\partial x_i}.$$

▶ **Remark 5.3** Recall that the product of two functions $u, v \in L^1_{loc}(\Omega)$ may not be locally summable. In accordance and in contrast to classical differentiability, the product of two weakly differentiable functions on Ω may not be weakly differentiable. However if these two functions are additionally bounded, then their product is weakly differentiable and obeys the usual product rule.

Lemma 5.2
Consider a sequence of functions $(u_n) \subset L^1_{loc}(\Omega)$ and assume that each u_n admits the weak derivative $v_n = \frac{\partial u_n}{\partial x_i}$ for some $1 \le i \le N$. If $u_n \to u$ and $v_n \to v$ in $L^1_{loc}(\Omega)$, then u admits the weak partial derivative for with respect to x_i and $v = \frac{\partial u}{\partial x_i}$.

Proof For every test function $\varphi \in C^\infty_c(\Omega)$ we calculate as follows using the definition of the weak partial derivative

$$\int_\Omega v(x)\,\varphi(x)\,dx = \lim_{n\to+\infty} \int_\Omega v_n(x)\,\varphi(x)\,dx = -\lim_{n\to+\infty} \int_\Omega u_n(x)\,\frac{\partial\varphi(x)}{\partial x_i}\,dx$$

$$= -\int_\Omega u(x)\,\frac{\partial\varphi(x)}{\partial x_i}\,dx$$

which proves the assertion.

▶ **Remark 5.4** The aforementioned result is to be compared with the following well-known fact: Let $u_n, u, v : [0, 1] \to \mathbb{R}$ for $n \in \mathbb{N}$. Assume that $u_n \in C^1([0, 1])$ and that $u'_n \rightrightarrows v$, $u_n \to u$ for $[0, 1]$. Then

$$u'(t) = v(t) \text{ for } t \in [0, 1].$$

When u'_n converges merely pointwise to v, then it may happen that the above formula does not hold. Indeed, for

$$u_n(t) = \frac{t}{1 + n \cdot t^2}$$

we have $u_n \rightrightarrows u := 0$ on $[0, 1]$ and we easily compute that $\left(u'_n\right)$ converges pointwise to

$$v(t) = \begin{cases} 1, & t = 0, \\ 0, & t \ne 0. \end{cases}$$

—

Exercise 5.10

Assume $u \in L^1_{loc}(\Omega)$. Show that if the first-order weak partial derivatives of u satisfy

$$\frac{\partial u(x)}{\partial x_i} = 0 \quad \text{for } i = 1, 2, \ldots, N \text{ and a.e. } x \in \Omega,$$

then u coincides a.e. with a constant function.

Now we can proceed to defining the space $H_0^1(\Omega)$ that will be used for consideration of the Dirichlet problem.

Definition 5.3 (Space $H^1(\Omega)$)

The Sobolev space $H^1(\Omega)$ is the space of all functions $u : \Omega \to \mathbb{R}$ integrable with square such that the weak gradient ∇u exists and belongs to $L^2(\Omega; \mathbb{R}^N)$. On $H^1(\Omega)$ we shall use the norm

$$\|u\|_{H^1(\Omega)} = \sqrt{\int_\Omega |u(x)|^2\,dx + \int_\Omega |\nabla u(x)|^2\,dx}$$

induced by the inner product

$$(u, v)_{H^1(\Omega)} = \int_\Omega u(x)\,v(x)\,dx + \int_\Omega \nabla u(x)\,\nabla v(x)\,dx.$$

The definition of the space $H^2(\Omega)$ is now obvious: we take such functions from $H^1(\Omega)$ whose weak partial derivatives are weakly differentiable and integrable with square.

Definition 5.4 (Space $H_0^1(\Omega)$)

The subspace $H_0^1(\Omega) \subseteq H^1(\Omega)$ is defined as the closure of the space of test functions $C_c^\infty(\Omega)$ in the norm of space $H^1(\Omega)$. More precisely, $u \in H_0^1(\Omega)$ if and only if there exists a sequence of functions $(u_n) \subset C_c^\infty(\Omega)$ such that

$$\|u - u_n\|_{H_0^1(\Omega)} \to 0.$$

The space $C^\infty(\Omega) \cap H_0^1(\Omega)$ is dense in $H_0^1(\Omega)$. The property that any function in $H_0^1(\Omega)$ can be approximated by functions whose derivatives exist in the classical sense is often useful in proofs which employ the density argument.

Exercise 5.11

Prove that $H^1(\Omega)$ and $H_0^1(\Omega)$ are Hilbert spaces. Hint: Use the idea of the proof of Lemma 5.2 in order to prove that these spaces are complete.

Exercise 5.12

Fix $\gamma > 0$ and consider the radially symmetric function

$$u(x) = |x|^{-\gamma} = \left(\sum_{i=1}^N x_i^2\right)^{-\gamma/2}, \quad 0 < |x| < 1. \tag{5.7}$$

Prove that $u \in C^1(\Omega\backslash\{0\})$ with the partial derivatives

$$u_{x_i} = -\frac{\gamma}{2}\left(\sum_{i=1}^{N} x_i^2\right)^{-(\gamma/2)-1} 2x_i = \frac{-\gamma x_i}{|x|^{\gamma+2}} \qquad (5.8)$$

and with the gradient $\nabla u = (u_{x_1}, \ldots, u_{x_N})$ outside of 0 having the following Euclidean norm a.e. on Ω

$$|\nabla u(x)| = \left(\sum_{i=1}^{N} |u_{x_i}(x)|^2\right)^{1/2} = \frac{\gamma}{|x|^{\gamma+1}}.$$

Exercise 5.13

Assume that $\gamma < \frac{N-2}{2}$. Denote by σ_N the $(N-1)$-dimensional measure of the unit sphere $\{x \in \mathbb{R}^N; |x| = 1\} \subset \mathbb{R}^N$. Show that

$$\int_{x \in \mathbb{R}^N, |x| < 1} |x|^{-2(\gamma+1)} dx = \sigma_N \int_0^1 r^{N-1} r^{-2(\gamma+1)} dr.$$

Example 5.7

We will show that if $\gamma < \frac{N-2}{2}$ then the function u defined by (5.7) belongs to $H_0^1(\Omega)$. Indeed, computing as in the aforementioned exercise we conclude that

$$\int_\Omega \left(|u|^2 + |\nabla u|^2\right) dx < \infty.$$

It remains to be shown that functions given by (5.8) stand for weak derivatives. We know by the properties of u that the classical derivatives and the weak derivatives coincide on $\Omega\backslash\{0\}$, and we need to show that these coincide on the whole Ω. For this purpose, consider any test function $\varphi \in C_c^\infty(\Omega)$. Fix $i \in \{1, \ldots, N\}$. For convenience, we extend φ to the entire space $\mathbb{R}^N$ by setting $\varphi(x) = 0$ for $x \notin \Omega$. Since $N \geq 2$, the x_i-axis has N-dimensional Lebesgue measure zero. An integration by part yields

$$\int_\Omega \frac{-\gamma x_i}{|x|^{\gamma+2}} \varphi(x) dx = \int_{\mathbb{R}^{N-1}\backslash\{0\}} \left(\int_\mathbb{R} \frac{-\gamma x_i}{|x|^{\gamma+2}} \varphi(x) dx_i\right) dx_1 \cdots dx_{i-1} dx_{i+1} \cdots dx_N =$$

$$- \int_{\mathbb{R}^{N-1}\backslash\{0\}} \left(\int_\mathbb{R} |x|^{-\gamma} \varphi_{x_i}(x) dx_i\right) dx_1 \cdots dx_{i-1} dx_{i+1} \cdots dx_N$$

$$= -\int_\Omega |x|^{-\gamma} \varphi_{x_i}(x) dx$$

completing the proof.

▶ **Remark 5.5** Observe that the computation in Example 5.7 relied on the fact that u is absolutely continuous (in fact, smooth) on a.e. line parallel to one of the coordinate axes. However, it must be stressed that there is no way to change u on a set of measure zero, so that it becomes continuous on the whole domain Ω.

Now we turn to showing that in this case we have the Poincaré Inequality as well:

Theorem 5.3   (Poincaré Inequality)
There is a constant $C_P > 0$ depending on Ω such that

$$\sqrt{\int_\Omega |u(x)|^2 \, dx} \le C_P \sqrt{\int_\Omega |\nabla u(x)|^2 \, dx} \text{ for all } u \in H_0^1(\Omega).$$

Proof For any fixed $u \in H_0^1(\Omega)$, we obtain what follows by a direct calculation involving integration by parts and next the Schwarz Inequality

$$\int_\Omega |u(x)|^2 \, dx = \int_\Omega 1 \cdot |u(x)|^2 \, dx = -\int_\Omega x_1 \cdot \frac{\partial}{\partial x_1} |u(x)|^2 \, dx =$$
$$-2 \int_\Omega x_1 \cdot \frac{\partial u(x)}{\partial x_1} u(x) \, dx \le 2 |\Omega| \sqrt{\int_\Omega |u(x)|^2 \, dx} \sqrt{\int_\Omega |\nabla u(x)|^2 \, dx},$$

since $|x_1| \le 1$.

▶ **Remark 5.4** Note that in our case $C_P \le 2\pi^2$. In case of *the Poincaré Inequality* for the space $H_0^1(0, 1)$ we have reached the best constant that is the number $\frac{1}{\pi}$ which is an inverse of the first eigenvalue of the Dirichlet (negative) Laplacian. Such a result, apart from what we have seen in Sect. 4.6, can be reached by minimizing over $H_0^1(0, 1)$ of a functional

$$u \mapsto \|u\|_{H_0^1}^2$$

with respect to the constraint

$$\|u\|_{L^2}^2 - 1.$$

This task can be worked out with the use of *the Lagrange Multiplier Rule* exactly as we did in Sect. 1.6 with several details being different due to the infinite dimensional setting. Nevertheless for higher-dimensional domains, it is not easy to reach the best constant. We refer for example to the recent paper [31] for some discussion about this issue. Let $N > 2$ and put

$$T = \frac{1}{\sqrt{\pi N \, (N-2)}} \left(\frac{\Gamma \left(1 + \frac{N}{2}\right) \Gamma(N)}{\Gamma \left(\frac{N}{2}\right) \Gamma \left(1 + \frac{N}{2}\right)} \right)^{\frac{1}{N}},$$

where Γ is the gamma function. Following [47] and denoting $T_q := T |\Omega|^{\frac{2^*-q}{2^*q}}$ we have the inequality

$$\|u\|_q \leq T_q \|u\|, \text{ for all } u \in \mathrm{H}_0^1(0, 1).$$

for all $q \in [1, 2^*)$. By 2^* we denote the Sobolev conjugate exponent for $N > 2$

$$2^* = \frac{2N}{N-2}.$$

Note that $2^* > 2$. In our case of course $T_q := T$.

Exercise 5.14

Show that by *the Poincaré Inequality* we have an equivalent norm on $\mathrm{H}_0^1(\Omega)$ given by

$$\|u\|_{\mathrm{H}_0^1(\Omega)} = \sqrt{\int_\Omega |\nabla u \, (x)|^2 \, dx} \text{ for } u \in \mathrm{H}_0^1(\Omega).$$

Exercise 5.15

Using *the Poincaré Inequality* , suggest the norm on the space $\mathrm{H}_0^1(\Omega) \cap \mathrm{H}^2(\Omega)$.

It remains to cover the case of continuous and compact imbedding of the space $\mathrm{H}_0^1(\Omega)$ into suitable spaces of integrable functions.

Theorem 5.4 (Imbeddings for $H_0^1(\Omega)$)

If $N > 2$, then $\mathrm{H}_0^1(\Omega) \overset{c}{\hookrightarrow} \mathrm{L}^q (\Omega)$ *for every $q \in [1, 2^*)$ and also* $\mathrm{H}_0^1(\Omega) \hookrightarrow \mathrm{L}^{2^*} (\Omega)$. *If $N = 2$ then* $\mathrm{H}_0^1(\Omega) \overset{c}{\hookrightarrow} \mathrm{L}^q (\Omega)$ *for every $q \in [1, \infty)$.*

5.4   Some Applications to Integral Functionals

Let us recall that $\Omega \subseteq \mathbb{R}^N$ is an open unit ball. We will distinguish between the cases $N = 2$ and $N > 2$ as far as some growth assumptions on the integrand are concerned in several applications. We proceed to showing how to manage counterparts of certain results obtained in Chaps. 3 and 4 to the case of functionals involving multiple integrals. We would like to underline at the very beginning that it is not possible to obtain the exact analogy of the regularity results connected to *the du Bois–Reymond Lemma*. The result that we can shift from the interval to higher-dimensional domains is as follows:

Lemma 5.3

Let $u \in C\left(\overline{\Omega}\right)$ be fixed. Assume that

$$\int_\Omega u(x) v(x) \, dx = 0$$

for all $v \in C^1\left(\overline{\Omega}\right)$ with the property that $v|_{\partial\Omega} = 0$. Then $u(x) = 0$ for $x \in \Omega$.

Proof Suppose there is some $c \in \Omega$ that $u(c) > 0$. Then there is some closed cube $[\alpha, \beta]^N \subset \Omega$ over which u is positive. Consider a $C^1\left(\overline{\Omega}\right)$ test function $v(x) = v_1(x_1) \cdot \ldots \cdot v_N(x_N)$, where

$$v_i(x_i) = \begin{cases} (x_i - \alpha)^2 (\beta - x_i)^2, & x_i \in [\alpha, \beta], \\ 0, & x_i \notin [\alpha, \beta] \end{cases}$$

for $1 \leq i \leq N$. We easily see that $v \in C^1\left(\overline{\Omega}\right)$ and has the property that $v|_{\partial\Omega} = 0$. Moreover $v(x) > 0$ for $x \in [\alpha, \beta]^N$ and $v \equiv 0$ otherwise. Therefore, we have

$$\int_\Omega u(x) v(x) \, dx > 0.$$

Hence we arrived at a contradiction.

Exercise 5.16

Formulate the higher-order version of Lemma 5.3 as a counterpart of Lemma 2.9.

Exercise 5.17

Formulate and prove the counterpart of Lemma 5.3 for $H_0^1(\Omega)$ test functions. What can we assume about function u in this case?

Careful examination of remaining proofs from Sect. 2.7 tells us that the other results work only for the one variable case. We can now proceed to some examples and applications to integral functionals defined on $H_0^1(\Omega)$.

Exercise 5.18

Let $\alpha, \beta > 0$. Assume that $f : \Omega \times \mathbb{R} \to \mathbb{R}$ is a Carathéodory function and that there exist $\varphi \in L^\infty(\Omega, \mathbb{R}_+), h \in L^\beta(\Omega)$ such that

$$|f(x, u)| \leq \varphi(x)|u|^\alpha + h(x) \tag{5.9}$$

for a.e. $x \in \Omega$ and all $u \in \mathbb{R}$. Find the values of parameters α, β for which the functional $F : L^2(\Omega) \to \mathbb{R}$ defined by

$$F(u) = \int_\Omega f(x, u(x))\, dx$$

is continuous. Hint: Employ *the Krasnosel'skii Theorem*.

Exercise 5.19

Find values of α, β for which the functional F given in the above exercise is weakly continuous when defined on $H_0^1(\Omega)$. Hint: Distinguish between cases $N = 2$ and $N > 2$ and use Theorem 5.4.

▶ **Remark 5.5** In Example 2.7 we considered an integral functional defined by the continuous convex integrand $G : \mathbb{R} \to \mathbb{R}$ and considered over $H_0^1(0, 1)$. It was well posed and weakly continuous there. When considered on $H_0^1(\Omega)$, the functional

$$F : H_0^1(\Omega) \to \mathbb{R}, \ F(u) = \int_\Omega G(u(x))\, dx$$

need not be well posed without any further growth conditions imposed on G, namely we can employ the conditions given in (5.9).

Now we turn to the calculation of the Gâteaux and Fréchet derivatives. We may use of the following counterpart of the Leibniz rule concerning differentiation under integral sign for multiple integrals which follows directly from the application of *the Lebesgue Dominated Convergence Theorem* and *the Mean Value Theorem*:

Theorem 5.5 (Leibniz rule in several space dimension)
Assume that both $f : \Omega \times \mathbb{R} \to \mathbb{R}$ and its partial derivative with respect to the second variable $\frac{\partial f}{\partial u} : \Omega \times \mathbb{R} \to \mathbb{R}$ are Carathéodory functions. Let there

---

(Continued)

Theorem 5.5 (continued)

exists un $L^1(\Omega)$ *function* ξ *such that*

$$\left|\frac{\partial}{\partial u} f(x, u)\right| \leq \xi(x) \text{ for a.e. } x \in \Omega \text{ and all } u \in \mathbb{R}.$$

Then

$$\frac{\partial}{\partial u} \int_\Omega f(x, u)\, dx = \int_\Omega \frac{\partial}{\partial u} f(x, u)\, dx \text{ for all } u \in \mathbb{R}.$$

Example 5.8

We start with rewriting Example 3.4. We fix $\rho \in C(\overline{\Omega})$, put $E = C^1(\overline{\Omega})$ and define $F : E \to \mathbb{R}$ as follows

$$F(u) = \int_\Omega \rho(x)\sqrt{1 + |\nabla u(x)|^2}\, dx. \tag{5.10}$$

We will proceed using the method described in Remark 3.2 in order to differentiate F. We note that

$$\frac{|\rho(x)z|}{\sqrt{1 + |z|^2}} \leq |\rho(x)| \text{ for all } x \in \Omega, \ z \in \mathbb{R}.$$

Therefore, we can apply Theorem 5.5, in order to obtain for fixed $u, v \in E$ and for $\varepsilon \in \mathbb{R}$ that

$$\frac{\partial}{\partial \varepsilon} F(u + \varepsilon v) = \int_\Omega \frac{\partial}{\partial \varepsilon}\left[\rho(x)\sqrt{1 + |\nabla u(x) + \varepsilon \nabla v(x)|^2}\right] dx =$$

$$\int_\Omega \frac{\rho(x)(\nabla u(x) + \varepsilon \nabla v(x))\nabla v(x)}{\sqrt{1 + |\nabla u(x) + \varepsilon \nabla v(x)|^2}}\, dx.$$

Evaluating at $\varepsilon = 0$ we have

$$\delta F(u; v) = \int_\Omega \frac{\rho(x)\nabla u(x)\nabla v(x)}{\sqrt{1 + |\nabla u(x)|^2}}\, dx.$$

By a direct calculation, we obtain that formula $v \mapsto \delta F(u; v)$ defines a linear and bounded functional on E. Indeed,

$$
|\delta F(u; v)| \leq \int_\Omega \frac{|\rho(x)\nabla u(x)|}{\sqrt{1 + |\nabla u(x)|^2}} dx \cdot \max_{x \in \overline{\Omega}} |\nabla v(x)|
$$

$$
\leq \int_\Omega \frac{|\rho(x)\nabla u(x)|}{\sqrt{1 + |\nabla u(x)|^2}} dx \cdot \|\nabla v\|_\infty \text{ for all } v \in E.
$$

Using estimation

$$
\left| \frac{x}{\sqrt{1 + x^2}} - \frac{y}{\sqrt{1 + y^2}} \right| \leq |x - y| \text{ for all } x, y \in \mathbb{R}
$$

we will show the continuity of $\delta F(\cdot; v)$ with respect to u uniform in v from the unit sphere in E. Indeed, given a sequence $(u_n) \subset E$ convergent to some u_0, we obtain for $v \in E$ with $\|v\|_\infty = 1$:

$$
\int_\Omega \left| \frac{\rho(x)\nabla u_n(x)}{\sqrt{1 + |\nabla u_n(x)|^2}} - \frac{\rho(x)\nabla u_0(x)}{\sqrt{1 + |\nabla u_0(x)|^2}} \right| |\nabla v(x)| \, dx \leq \|\nabla u_n - \nabla u_0\|_\infty \|v\|_\infty .
$$

Therefore, the functional considered is continuously differentiable.

Example 5.9

Note that a difference arises when performing some calculations due to the fact that functions from $H_0^1(\Omega)$ need not be continuous in contrast to functions from $H_0^1(0, 1)$. We show how to deal with such a situation by examining the counterpart of Example 3.8. Due to the modified setting we will need to adjust both the calculations and the assumptions. We put $E = H_0^1(\Omega)$ and fix $q \in (1, 2^*)$. Let $g : \mathbb{R} \to \mathbb{R}$ be a C^1 function such that there are constants $a, b > 0$

$$
|g_u(u)| \leq a + b |u|^{q-1} \text{ for all } u \in \mathbb{R}. \tag{5.11}
$$

By a direct calculation we find numbers $a_1, b_1, c_1 > 0$ such that

$$
|g(u)| \leq c_1 + b_1 |u| + a_1 |u|^q \text{ for all } u \in \mathbb{R}.
$$

We consider functional $F : E \to \mathbb{R}$ given by

$$
F(x) = \int_\Omega g(u(x)) \, dx + \frac{1}{2} \int_\Omega |\nabla u(x)|^2 \, dx
$$

which is well defined on E due to the above given estimation and due to Theorem 5.4. Using Example 3.12 we see that the second term is obviously C^1.

Following Example 3.8 we suppose F has the Gâteaux derivative at every point u and in each direction h defined by the formula

$$\left\langle F'(u), h\right\rangle = \int_\Omega g_u\left(u(x)\right)h\left(x\right)dx + \int_\Omega \nabla u\left(x\right)\nabla h\left(x\right)dx.$$

Indeed, as in the real line case, the application of the Lagrange Mean Value Theorem says that

$$\left|\frac{g(z+\varepsilon w)-g(z)}{\varepsilon}\right| \leq \max_{c\in[z,w]}|g_u(c)|\,|w| \text{ for any } z, w \in \mathbb{R} \text{ and any } \varepsilon > 0.$$

By the above estimation together with assumption (5.11), we can apply *the Lebesgue Dominated Convergence Theorem* in order to differentiate under the integral sign. We check that

$$h \to \int_\Omega g_u\left(u(x)\right)h\left(x\right)dx$$

defines a linear bounded functional on E. Indeed, by *the Schwarz Inequality* and further *the Poincaré Inequality* we have

$$\int_\Omega |g_u\left(u(x)\right)h\left(x\right)|dx \leq \sqrt{\int_\Omega |g_u\left(u(x)\right)|^2dx}\sqrt{\int_\Omega |h\left(x\right)|^2dx} \leq g_1\|h\|^2_{H^1_0(\Omega)},$$

where $g_1 = \sqrt{\int_\Omega |g_u\left(u(x)\right)|^2\,dx}$ is a finite number due to (5.11). Thus we reach the conclusion that F is Gâteaux differentiable over E.

Exercise 5.20

Assume that $E = H^1_0(\Omega)$, let $\rho \in L^2\left(\Omega\right)$ and consider functional F given by formula (5.10). Prove that it is well posed and next that F is continuously differentiable.

Exercise 5.21

Check whether the functional F defined in Example 5.9 is continuously differentiable.

Exercise 5.22

Let $f \in C\left(\overline{\Omega} \times \mathbb{R}^2\right)$. Prove that the functional

$$F(u) = \int_\Omega f\left(x, u(x), \nabla u\left(x\right)\right)dx$$

is well defined over $E = C^1(\overline{\Omega})$. Under additional assumptions that the partial derivatives f_u, f_z of f with respect to second and third variable exist and are jointly continuous prove that F is C^1 and that

$$\left\langle F'(u), h \right\rangle = \int_{\Omega} \left(f_u(x, u(x), \nabla u(x)) h(x) + f_z(x, u(x), \nabla u(x)) \nabla h(x) \right) dx,$$

for all $u, h \in C^1(\overline{\Omega})$. Consider also the special case of functional

$$F(u) = \int_{\Omega} g(x, u(x)) \, dx + \frac{1}{2} \int_{\Omega} |\nabla u(x)|^2 \, dx$$

under suitable (which?) assumptions on g.

Example 5.10

In Sect. 4.3, we considered minimization of a classical action functional. As a counterpart of it for multiple integrals, we consider $J : H_0^1(\Omega) \to \mathbb{R}$ given by

$$J(u) = \frac{1}{2} \int_{\Omega} |\nabla u(x)|^2 \, dx + \frac{1}{\alpha} \int_{\Omega} |u(x)|^\alpha \, dx + \int_{\Omega} g(x) u(x) \, dx,$$

where $2 \le \alpha < 2^*$ and we recall that 2^* is the Sobolev conjugate exponent. Here $g \in L^2(\Omega)$. In case $N = 2$ we take any $\alpha > 1$. We easily show that J is C^1 over $H_0^1(\Omega)$ and that

$$\left\langle F'(u), h \right\rangle = \int_{\Omega} \left(\nabla u(x) \nabla h(x) + |u(x)|^{\alpha-2} u(x) h(x) + g(x) u(x) \right) dx$$

for all $u, h \in H_0^1(\Omega)$. Note that in case of the functional defined in Sect. 4.3, we can take any exponent because of $H_0^1(0, 1)$ into the space of the absolutely continuous functions.

Exercise 5.23

Let $N > 2$. Show that when $\alpha = 2$ the functional J from Example 5.10 is strictly convex. Show that for any $2 \le \alpha \le 2^*$ the functional J is coercive. Formulate the counterpart of this result for $N = 2$.

Exercise 5.24

Let $N > 2$. Let $g : \mathbb{R} \to \mathbb{R}$ be a continuously differentiable convex function with a derivative $g_u : \mathbb{R} \to \mathbb{R}$ and consider the functional $J : H_0^1(\Omega) \to \mathbb{R}$ given by

$$J(u) = \frac{1}{2} \int_{\Omega} |\nabla u(x)|^2 \, dx + \int_{\Omega} g(u(x)) \, dx.$$

Assume that there are constants $a, b > 0$ such that

$$|g_u(u)| \leq a + b|u|^{2^*-1} \text{ for all } u \in \mathbb{R}.$$

Prove that J is well defined, differentiable in the sense of Gâteaux, sequentially weakly lower semicontinuous and coercive. Argue that J has exactly one minimizer. Note that we can allow the growth to be at most of order $2^* - 1$ since now we require only continuous embedding and not the compact one.

5.5   On the Dirichlet Problem

By $C_0(\overline{\Omega})$ we mean the space of such functions continuous defined on $\overline{\Omega}$ that vanish on the boundary. When $u \in C^2(\overline{\Omega})$ we can define the following quantity known as the (negative) Laplacian

$$-\Delta u(x) := -\sum_{i=1}^{N} \frac{\partial^2 u(x)}{\partial x_i^2} \text{ for } x \in \Omega$$

and we can consider the associated Poisson problem

$$-\Delta u(x) = f(x) \text{ for } x \in \Omega, \ u|_{\partial\Omega} = 0 \tag{5.12}$$

for some fixed continuous function $f : \overline{\Omega} \to \mathbb{R}$. We then look for classical solutions of (5.12), i.e. in the space $C_0(\overline{\Omega}) \cap C^2(\overline{\Omega})$. By a direct calculation we show that the classical solutions to (5.12) are critical points to the following strictly convex functional

$$J(u) = \frac{1}{2} \int_\Omega |\nabla u(x)|^2 \, dx - \int_\Omega f(x) u(x) \, dx. \tag{5.13}$$

Exercise 5.25

Prove that functional J given by (5.13) is continuously differentiable on both $C_0(\overline{\Omega}) \cap C^2(\overline{\Omega})$ and $H_0^1(\Omega)$. Prove that J is coercive and sequentially weakly lower semicontinuous on $H_0^1(\Omega)$. Modify the assumptions on f in case J is considered over $H_0^1(\Omega)$.

However, again having fixed $u \in C_0(\overline{\Omega}) \cap C^2(\overline{\Omega})$, we can easily calculate that

$$\int_{\Omega} \left(-\sum_{i=1}^{N} \frac{\partial^2 u(x)}{\partial x_i^2} \right) v(x)\, dx = \int_{\Omega} \nabla u(x)\, \nabla v(x)\, dx \qquad (5.14)$$

for any $v \in C_0(\overline{\Omega}) \cap C^1(\overline{\Omega})$. We see that for the right-hand side of (5.14) to be well defined, it suffice to have $u, v \in H^1(\Omega)$.

Exercise 5.26

Define an operator

$$-\Delta : H_0^1(\Omega) \to \left(H_0^1(\Omega) \right)^*$$

by the formula

$$\langle -\Delta u, v \rangle = \int_{\Omega} \nabla u(x)\, \nabla v(x)\, dx \text{ for } u, v \in H_0^1(\Omega).$$

Prove that $-\Delta$ is linear and continuous. Observe that as in Remark 4.5 operator $-\Delta$ sends a point from $H_0^1(\Omega)$ (namely function u) into a continuous and linear functional acting on $H_0^1(\Omega)$, i.e., an element of $\left(H_0^1(\Omega) \right)^*$. We call the operator obtained the (negative) Laplacian.

Using the outcome of the aforementioned exercise, we introduce the definition of a bilinear form $a : H_0^1(\Omega) \to \mathbb{R}$ by

$$a(u, v) = \int_{\Omega} \nabla u(x)\, \nabla v(x)\, dx.$$

Applying *the Lax–Milgram Theorem* (see Theorem 4.9) we see that for any $f \in L^2(\Omega)$ there is exactly one $u_0 \in H_0^1(\Omega)$ satisfying

$$\int_{\Omega} \nabla u_0(x)\, \nabla v(x)\, dx = \int_{\Omega} f(x)\, v(x)\, dx \text{ for all } v \in H_0^1(\Omega)$$

which minimizes functional J given by (5.13). We call such an element u_0 a weak solution, or a solution in the sense of space $H_0^1(\Omega)$ to the Dirichlet problem (5.12). We shall denote this problem by

$$-\Delta u(x) = f(x) \text{ for } x \in \Omega, \ u|_{\partial \Omega} = 0$$

provided that it is understood in the weak sense described earlier.

▶ **Remark 5.6** It is not easy to obtain some further information on the regularity of solutions to (5.12). This is known as the Liebermann regularity theory, see [36]. We will not however proceed further with it. On the other hand, some $H^2(\Omega)$ regularity theory is contained in [18] which we will now follow. Theorem 1 from Section 8.3 in [18] says that if $f \in L^2(\Omega)$ and $u_0 \in H_0^1(\Omega)$ is a weak solution to (5.12) then $u_0 \in H^2(\Omega)$. Note, however, the special type of domain that we consider here.

From now on we consider the following Dirichlet Problem (understood in a weak sense): find a function $u \in H_0^1(\Omega)$ such that the following equation is satisfied:

$$\begin{cases} -\Delta u(x) + g(x, u(x)) = 0, \\ \quad u|_{\partial\Omega} = 0 \end{cases} \tag{5.15}$$

under the assumption that $N > 2$ and that g satisfies:

B1 $g : \Omega \times \mathbb{R} \to \mathbb{R}$ *is a Carathéodory function for which there is a function* $a \in L^2(\Omega)$ *and constant* $b > 0$ *and* $q \in (1, 2^*)$ *that*

$$|g(x, u)| \le a + b|u|^{q-1} \text{ for all } u \in \mathbb{R} \text{ and a.e. } x \in \Omega. \tag{5.16}$$

We say that $u \in H_0^1(\Omega)$ is a weak solution to 5.15) provided that

$$\int_\Omega \nabla u(x) \nabla v(x) \, dx + \int_\Omega g(x, u(x)) v(x) \, dx = 0 \text{ for all } v \in H_0^1(\Omega). \tag{5.17}$$

Exercise 5.27

Assume that condition **B1** is satisfied. Show that the formula (5.17) is well defined.

We define $G : \Omega \times \mathbb{R} \to \mathbb{R}$ by

$$G(x, u) = \int_0^u g(x, s) \, ds \text{ for a.e. } x \in \Omega \text{ and all } x \in u. \tag{5.18}$$

Exercise 5.28

Assume that condition **B1** is satisfied. Prove that G defined by (5.18) is a Carathéodory function as well and find the relevant growth conditions imposed on G.

The corresponding Euler action functional $J : H_0^1(\Omega) \to \mathbb{R}$ (i.e. a functional whose critical points are weak solutions to (5.15)) is given by

$$J(u) = \frac{1}{2} \int_\Omega |\nabla u(x)|^2 \, dx + \int_\Omega G(x, u(x)) \, dx. \qquad (5.19)$$

Exercise 5.29

Assume that condition **B1** is satisfied. Prove that J given by (5.19) is well defined (i.e. finite for every $u \in H_0^1(\Omega)$). Show that J is C^1 and sequentially weakly lower semicontinuous over $H_0^1(\Omega)$.

In order to be able to apply Theorem 4.3 we need to assume a type of a growth condition on G which leads to the coercivity of J, namely:

B2 *there exist functions* $a \in L^\infty(\Omega)$, $b, c \in L^2(\Omega)$ *and a number* $\beta \in (0, 2)$ *such that for a.e.* $x \in \Omega$ *and all* $u \in \mathbb{R}$ *we have*

$$G(x, u) \geq -\frac{1}{2} a(x) |u|^\beta + b(x) u + c(x).$$

Exercise 5.30

Assume that condition **B1, B2** are satisfied. Prove that J given by (5.19) is coercive.

Exercise 5.31

Find values of parameters $\alpha, \beta, a > 0$ for which functional J is coercive in case

$$G(x, u) \geq -a |u|^\beta + b(x) |u|^\alpha$$

for a.e. $x \in \Omega$ and all $u \in \mathbb{R}$.

With the aforementioned preparation, we can easily prove the following existence result:

Theorem 5.6
*Assume that conditions **B1, B2** are satisfied. Then the Dirichlet Problem (5.15) has at least one weak solution* $u_0 \in H_0^1(\Omega) \cap H^2(\Omega)$.

Proof Under conditions **B1, B2** the functional J given by (5.19) is coercive, sequentially weakly lower semicontinuous and continuously differentiable. Hence functional J has at least one critical point $u_0 \in H^1_0(\Omega)$. By Remark 5.6 we see that $u_0 \in H^1_0(\Omega) \cap H^2(\Omega)$ since the right-hand side of (5.15) belongs to $L^2(\Omega)$.

▶ **Remark 5.7** In case when G is convex we can alter condition (5.16) in assumption **B1** as follows:

$$|g(x, u)| \leq a + b|u|^{2^*-1} \text{ for all } u \in \mathbb{R} \text{ and } a.e. \ x \in \Omega.$$

This due to the fact that upon convexity in order to prove that the functional J is sequentially weakly lower semicontinuous it suffices to demonstrate that it is well defined and continuous. Hence, the compact embedding is not utilized.

Exercise 5.32

Provide a detailed proof of Theorem 5.6 and of assertion from Remark 5.7.

Exercise 5.33

Formulate conditions on the nonlinear term G, which lead to the existence and uniqueness result.

Exercise 5.34

Formulate assumptions on relevant functions in condition

$$G(x, u) \geq -\frac{1}{2}a(x)|u|^2 + b(x)u + c(x)$$

leading to the coercivity of J. Provide a suitable version of Theorem 5.6 in this case.

▶ **Remark 5.8** It remains to comment on what happens with the Dirichlet Problem when $N = 2$. In this case we need to adjust the growth condition in assumption **B1** by replacing exponent $q - 1$ with any $r \in (1, +\infty)$. The other conditions are left unchanged here.

References

These notes grew out from reading several background books, mainly [32] and [49] coupled with [37]. For the theory of convex functions, we suggest [17]. The preliminary Polish version of this text is contained in [26] with some additional material on the one hand and with different presentation and approach on the other. The book [26] is centered around the Dirichlet problem and not as we do on the Weierstrass Theorem. We also used a lot of inspiration from [16] and [20] that are very advanced texts. As the functional analysis background we used mainly [8] and [28], while for the Lebesgue integration [48]. As for related results on the application of the theory of monotone operators that corresponds with these notes, we suggest [22] and [26]. There exists a number of advanced texts which we suggest to the Reader in order to extend their knowledge in the variational methods: [19, 33, 39, 40, 52]. Moreover, we have included in the references other sources that perhaps are not cited explicitly, are related to the topic of this text, and could serve as suggested further reading. The reader may wish to check [50] for convex optimization problems in a general Banach space under the presence of computational errors. While we exploit compactness in our approach, in [51] approximate solutions to minimization problems are investigated via some generic principles. A lot of optimization tools is to be found in [13].

1. R.A. Adams, *Sobolev Spaces* (Academic Press, New York, 1975)
2. A. Ambrosetti, G. Prodi, *A Primer of Nonlinear Analysis.* Cambridge Studies in Advanced Mathematics, vol. 34 (Cambridge University Press, Cambridge, 1995), 180 p.
3. R.P. Agarwal, *Difference Equations and Inequalities: Theory, Methods and Applications* (Marcel Dekker, New York, 2000)
4. V.M. Alekseev, V.M. Tikhomirov, S.V. Fomin, *Optimal control. Transl. from the Russian by V. M. Volosov.* Contemporary Soviet Mathematics (Consultants Bureau, New York, 1987)
5. M.S. Bazaraa, H.D. Sherali, C.M. Shetty, *Nonlinear Programming. Theory and Algorithms* (John Wiley & Sons, Hoboken, 2006)
6. M. Bełdziński, M. Galewski, Global diffeomorphism theorem applied to the solvability of discrete and continuous boundary value problems. J. Differ. Equ. Appl. **24**(2), 277–290 (2018)
7. A. Bressan, *Lecture Notes on Functional Analysis. With Applications to Linear Partial Differential Equations.* Graduate Studies in Mathematics, vol. 143 (American Mathematical Society, Providence, 2013)
8. H. Brézis, *Functional Analysis, Sobolev Spaces and Partial Differential Equations* (Springer, Berlin, 2010)

© The Author(s), under exclusive license to Springer Nature Switzerland AG 2024
M. Galewski, *Basics of Nonlinear Optimization*, Compact Textbooks in Mathematics, https://doi.org/10.1007/978-3-031-77160-6

9. J. Brinkhuis, V. Protasov, A new proof of the Lagrange multiplier rule. Oper. Res. Lett. **44**(3), 400–402 (2016)
10. D. Bucur, G. Buttazzo, *Variational Methods in Shape Optimization Problems* (Birkhauser, Boston, 2005)
11. C. Canuto, A. Tabacco, *Mathematical Analysis I* (Springer, Berlin, 2008)
12. C. Canuto, A. Tabacco, *Mathematical Analysis II* (Springer; Berlin, 2008)
13. L. Cesari, *Optimization, Theory and Applications* (Springer, New York, 1983)
14. F.H. Clarke,*Optimization and Nonsmooth Analysis.* Classics in Applied Mathematics, vol. 5 (Society for Industrial and Applied Mathematics, Philadelphia, 1990)
15. O. Dovgoshey, O. Martio, V. Ryazanov, M. Vuorinen, The Cantor function. Expo. Math. **24**, 1–37 (2006)
16. P. Drábek, J. Milota, *Methods of Nonlinear Analysis. Applications to Differential Equations.* Birkhäuser Advanced Texts. Basler Lehrbücher (Birkhäuser, Basel, 2007)
17. I. Ekeland, R. Temam, *Convex Analysis and Variational Problems* (North-Holland, Amsterdam, 1976)
18. L.C. Evans, *Partial Differential Equations. Graduate Studies in Mathematics*, vol. 19 (American Mathematical Society, Providence, 1998)
19. D.G. Figueredo, *Lectures on the Ekeland Variational Principle with Applications and Detours.* Preliminary Lecture Notes (SISSA, 1988)
20. S. Fučik, A. Kufner, *Nonlinear Differential Equations.* Studies in Applied Mechanics, vol. 2 (Elsevier, Amsterdam, Oxford, New York, 1980)
21. R. Gaines, *Difference equations associated with boundary value problems for second order nonlinear ordinary differential equations.* SIAM J. Numer. Anal. **11**, 411–434 (1974)
22. H. Gajewski, K. Gröger, K. Zacharias, *Nichtlineare Operatorgleichungen und Operatordifferentialgleichungen* (Akademie-Verlag, Berlin, 1974)
23. Eh. M. Galeev, V.M. Tikhomirov, *A Short Course on the Theory of Extremal Problems. Textbook.* (in Russian) (Moskva, 1989)
24. M. Galewski, E. Schmeidel, Non-spurious solutions to discrete boundary value problems through variational methods. J. Differ. Equ. Appl. **21**(12), 1234–1243 (2015)
25. M. Galewski, M. Rădulescu, On a global implicit function theorem for locally Lipschitz maps via non-smooth critical point theory. Quaest. Math. **41**(4), 515–528 (2018)
26. M. Galewski, *Wprowadzenie do metod wariacyjnych* (Wydawnictwo Politechniki Łódzkiej, Łódź, 2020). ISBN 978-83-66287-37-2
27. M. Galewski, *Basic Monotonicity Methods with Some Applications.* Compact Textbooks in Mathematics (Birkhäuser, Basel; SpringerNature, Switzerland; Basingstoke, 2021). ISBN: 978-3-030-75308-5
28. M. Haase, *Functional Analysis. An Elementary Introduction.* Graduate Studies in Mathematics, vol. 156 (AMS, Providence, 2014)
29. G. Haeser, D. Oliveira dos Santos, *A Simple Proof of Existence of Lagrange Multipliers* (2024). arXiv:2402.05335
30. A.D. Ioffe, V.M. Tikhomirov, *Theory of Extremal Problems.* Studies in Mathematics and Its Applications, vol. 6 (North-Holland Publishing Company, Amsterdam, New York, Oxford, 1979)
31. N. Ioku, Attainability of the best Sobolev constant in a ball. Math. Ann. **375**(1–2), 1–16 (2019)
32. J. Jahn, *Introduction to the Theory of Nonlinear Optimization*, 3rd edn. (Springer, Berlin, 2007)
33. Y. Jabri, *The Mountain Pass Theorem. Variants, Generalizations and Some Applications.* Encyclopedia of Mathematics and its Applications, vol. 95 (Cambridge University Press, Cambridge, 2003)
34. W.G. Kelley, A.C. Peterson, *Difference Equations. An Introduction with Applications*, 2nd edn. (Harcourt/Academic Press, San Diego, 2001)
35. U. Ledzewicz, H. Schättler, *Geometric Optimal Control. Theory, Methods and Examples.* Interdisciplinary Applied Mathematics, vol. 38 (Springer, New York, 2012), xix, 640 p. ISBN 978-1-4614-3833-5/hbk; 978-1-4614-3834-2/ebook

36. G. Lieberman, The natural generalization of the natural conditions of Ladyzhenskaya and Uraltseva for elliptic equations. Comm. Partial Diff. Equ. **16**, 311–361 (1991)
37. J. Mahwin, *Problemes de Dirichlet Variationnels Non Linéaires*. Séminaire de Mathématiques Supérieures 104 (Montreal, 1987)
38. G. Molica Bisci, D.D. Repovš, On some variational algebraic problems. Adv. Nonlinear Anal. **2**(2), 127–146 (2013)
39. D. Motreanu, V.D. Rădulescu, *Variational and Nonvariational Methods in Nonlinear Analysis and Boundary Value Problems, Nonconvex Optimization and Its Applications* (Springer, Berlin, 2003)
40. N.S. Papageorgiou, V.D. Rădulescu, D.D. Repovš, *Nonlinear Analysis – Theory and Methods*. Springer Monographs in Mathematics (Springer, Cham, 2019)
41. A.L. Peressini, F.E. Sullivan, J.J. Uhl, *The Mathematics of Nonlinear Programming*. Undergraduate Texts in Mathematics (Springer, New York, 1988)
42. L.M. Perko, *Differential Equations and Dynamical Systems*. Texts in Applied Mathematics, vol. 7 (Springer, New York, 1991)
43. M. Renardy, R.C. Rogers, *An Introduction to Partial Differential Equations*, 2nd edn. Texts in applied mathematics, vol. 13 (Springer, New York, Berlin, Heidelberg, 2004)
44. E. O. Roxin, *Control Theory and Its Applications*. Stability and Control: Theory, Methods and Applications, vol. 4 (Gordon and Breach, Amsterdam, 1997), xvii, 180 p.
45. W. Rudin, *Principles of Mathematical Analysis*, 2nd edn. (McGraw-Hill Book, New York, 1964)
46. W. Rudin, *Functional analysis. McGraw-Hill Series in Higher Mathematics* (McGraw-Hill Book, New York-Düsseldorf-Johannesburg, 1973)
47. G. Talenti, Best constants in Sobolev inequality. Ann. Mat. Pura Appl. **110**, 353–372 (1976)
48. T. Tao, *An Introduction to Measure Theory*. Graduate Studies in Mathematics, vol. 126 (2011)
49. J.L. Troutman, *Variational Calculus and Optimal Control*. Optimization with Elementary Convexity, Undergraduate Texts in Mathematics (Springer, New York, 1996)
50. A.J. Zaslavski, *Optimization in Banach Spaces*. SpringerBriefs in Optimization (Springer, Cham, 2022)
51. A.J. Zaslavski, *Optimization on Metric and Normed Spaces*. Springer Optimization and Its Applications, vol. 44 (Springer, New York, 2010)
52. E. Zeidler, *Applied Functional Analysis. Main Principles and Their Applications*. Applied Mathematical Sciences, vol. 109 (Springer, New York, 1995), xvi, 404 p.

Index

© The Author(s), under exclusive license to Springer Nature Switzerland AG 2024
M. Galewski, *Basics of Nonlinear Optimization*, Compact Textbooks in
Mathematics, https://doi.org/10.1007/978-3-031-77160-6